ACPL ITEM DISCARDED

```
*                        7029687
621.47
B24j
Barling, John
John Barling's Solar fun book
```

DO NOT REMOVE CARDS FROM POCKET

ALLEN COUNTY PUBLIC LIBRARY

FORT WAYNE, INDIANA 46802

You may return this book to any agency, branch, or bookmobile of the Allen County Public Library.

John Barling's Solar Fun Book

18 Projects for the Weekend Builder

621.47
B24j
Barling, John
John Barling's Solar fun book

7029687

**DO NOT REMOVE
CARDS FROM POCKET**

ALLEN COUNTY PUBLIC LIBRARY

FORT WAYNE, INDIANA 46802

You may return this book to any agency, branch,
or bookmobile of the Allen County Public Library.

DEMCO

John Barling's Solar Fun Book
18 Projects for the Weekend Builder

John Barling's Solar Fun Book
18 Projects for the Weekend Builder

John Barling

BRICK HOUSE PUBLISHING COMPANY
Andover, Massachusetts

ALLEN COUNTY PUBLIC LIBRARY
FORT WAYNE, INDIANA

Published by   Brick House Publishing Company
3 Main Street
Andover Massachusetts 01810

Production credits   Editor: Jack Howell
Copy editor: John Woodman
Book design: Herb Caswell
Cover design: Beth Anderson
Project drawings: Arthur Wheatley
Typesetting: Foam House Composition

We recommend care and adherence to standard construction safety procedures. Use welding glasses or the equivalent to protect your eyes when working with concentrating devices and when using power tools. Neither the publisher nor the author takes responsibility for accidents that may occur during the building or use of any of the solar projects described in this book.

Printed in the United States of America

Copyright © 1979 by John Barling.

All rights reserved. No part of this book may be reproduced in any form without the written permission of the publisher.

Library of Congress Cataloging in Publication Data

Barling, John.
  John Barling's Solar fun book.

  Bibliography: p.
  1. Solar energy. I. Title. II. Title: Solar fun book.
TJ810.B37   621.47   79-51371
ISBN 0-931790-11-5
ISBN 0-931790-04-2 pbk.

Acknowledgments

Many thanks to B.C.* Hydro, the Educational Research Institute of B.C., School District #21, B. C., and Greyhound Lines of Canada for their generous financial assistance, which enabled the development of so many projects.

The designs incorporated into this book would not have become a reality without the enthusiastic help of many individuals. I would particularly like to thank Erma Krebbers for the original drawings upon which the fine artwork of Art Wheatley is based, and Brenda Lukens, Elaine Brown, Nancy Conner, Raynier Pipke, Lorne Harrison, Marvin Kliewer, and John Ross for their devoted efforts in fully developing the original project concepts.

Thanks are due to Gaydene Giesbrecht for patiently typing and retyping so much material for the book. Last but not least a sincere "thank you" to my dear wife Bev for proofreading the text and for encouraging my interest in solar energy.

* British Columbia

Foreword

At a time when more and more people are coming to realize the sun's potential as a source of free heat, there has developed a real need for a simple do-it-yourself guide that gets away from complex formulas and endless complicated projections and gets back to basics in what is, in essence, a simple technology. This is just such a book. Its emphasis is on practical applications—a book that will help convey the tremendous recreative and money-saving possibilities of using solar energy.

Within these pages you will find a guide that demonstrates simply and effectively how the sun's heat can be used in a number of fascinating and effective ways by anyone with a straightforward handyman's approach and some basic tools. It's a book that will open up the mind and stimulate independent thinking at a time when all we are ever asked to do is to plug in and pay up; when the individual is fast becoming enslaved to so-called public service facilities and our sense of self-awareness and independence is being rapidly eroded.

It's gratifying to see ingenuity and conservation coming to the fore at this time, and there are probably few people better qualified than John Barling to write this book, covering as it does a wide range of solar-oriented projects. As a teacher, author and lecturer he has guided many along the fascinating path of solar exploration, and those who, like me, have had the pleasure of witnessing his work with students cannot help but be impressed by the enthusiasm and interest he generates among them with this exciting new technology.

A solid scientific base and many years of practical experience bounded by tight budgetary restrictions have made the author a master of economy and careful improvisation. His emphasis on the use of recycled materials wherever possible is in the best tradition of the "New Age" philosophy that recognizes that in an energy-hungry world we cannot go on raping the planet of its natural resources as we have done in the past and that in the sun's rays we have the only inexhaustible, nonpolluting, terrorist-resistant and free energy source.

Whatever your motivation for buying this book, whether you are concerned with the growing cost of conventional fuels and the toll on the ecological system of misguided energy policies or are just interested in "going solar" because it makes sound sense and you enjoy home projects, this rich collection of practical solar ideas will offer you fun in the sun. There is certainly no better motivation than that!

Dennis Milligan
President, Solair Systems Inc.
Vernon, B.C., Canada

Contents

Introduction 11

Cooking Devices

Hot Dog Cooker 15
Super Hot Dog Cooker 21
Solar Oven 25
Chicken Cooker 31
Parabolic Concentrating Cooker 37

Water and Air Heaters

Water Preheating Panel 43
Demonstration Hot Water Heater 47
Barling's Barrel Bread-Box Heater 51
Solar Swimming Pool Heater 55
Low-Cost Air Heater 69

Food-Producing Devices

Solar Cold Frame 75
Model Solar Greenhouse 79
Food Dryer 83

Experimental Solar Devices

Experimental Concentrating Water Heater 91
Fresnel Lens Concentrator 97
Giant Experimental Reflector 101

Miscellaneous Solar Devices

Glass Solar Still 109
Solar Wood Igniter 115
Bibliography 119

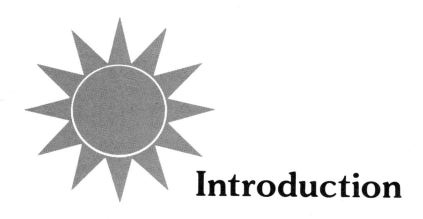

Introduction

One of the most exciting aspects of solar energy is the opportunity to develop different and unique ways of tapping the sun's power. There are so many possibilities for individuals to make contributions towards solar progress because developments still rely heavily on the contributions of the innovators and experimenters. My own personal involvement in developing designs and supervising their construction has truly been a tremendous experience; not only that, but harnessing solar energy is fun!

The projects were originally built by high school students working from plans that have since been updated and revised to produce clear, concise instructions that any handy person can follow.

Many people still view solar energy as a privilege of a rich minority of enthusiasts who enjoy exotic solar heating and cooling systems. In actual fact, anyone can get involved in solar by the construction of useful projects.

The projects in this book provide ideal material for adults and students alike to "switch on" to solar.

I am convinced that, if presented with the right experiences, the young people of today will form the less utility-dependent solar generation of tomorrow.

JOHN BARLING

Cooking Devices

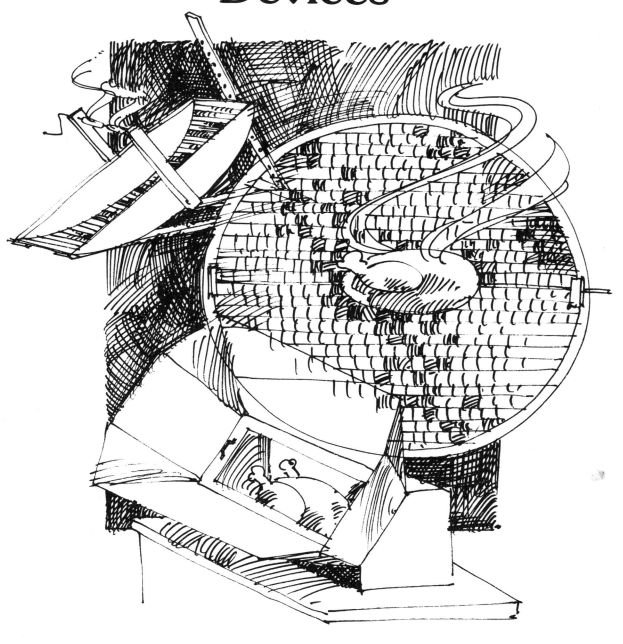

Hot Dog Cooker

The hot dog cooker is a project that is not difficult for anyone to build. It can be used to produce very good results, providing you have bright sunny weather. Bear in mind that any concentrating collector of this type will not work effectively with diffuse sunlight.

Three hot dogs or sausages can be cooked in about 3 or 4 minutes under ideal conditions. When weather conditions are less than ideal, wrapping the hot dogs with blackened foil or smoking them will cook them faster. Shish kebab or pieces of chicken can also be cooked. The dark color of the beef helps light absorption, and barbecue sauce on chicken will produce a mouth-watering treat!

This is a great project for a child-parent team to construct and helps quite dramatically to illustrate the real potential of sun power.

To draw a parabola

    1 piece of rigid cardboard from a large appliance carton

    3 foot length of 1-by-6-inch softwood

To construct the cooker

    1 4-by-8-foot sheet of ¾-inch plywood

    1 13⅝-by-51¼-inch piece of ⅛-inch hardboard

    5 1-foot-square mirror tiles

    1 small tube of clear silicone sealant

3 1½-inch tight-pinned butt hinges with screws

½ pound of 1¼-inch ring nails

waterproof glue

paint

glass cutter

Parabolic Reflectors

The reflecting surface of concentrating-type collectors such as this hot dog cooker is normally parabolic in shape. This shape produces a sharp focus of 1 inch or so.

Parabolic shapes are quite easily drawn. The specific method described can be adapted for any parabola you may wish to draw. Two things have to be decided: (a) how far away from the reflecting surface you wish the food to be placed, i.e. the focal point (in this case 18 inches), and (b) how long you want your parabola to be (in this case 4 feet).

Drawing a Parabola

1. Cut a sheet of rigid cardboard to produce a rectangle 30 by 30 inches (each corner must be absolutely square).

2. Using thumbtacks, firmly attach the card to an underlying sheet of plywood that is larger than the piece of cardboard.

3. Make a mark on one edge of the cardboard 18 inches from one corner. Also mark off one side of the cardboard in inches (see illustration).

4. Hammer a thin nail into the plywood next to the 18-inch mark on the cardboard.

5. Take a 3-foot-long board of 1-by-6-inch material that has absolutely square ends. Place this board with one long edge up against the nail.

6. Rotate the board to the right inch by inch and use a pencil to mark straight lines along the end of the wood as you rotate it. Ensure that the left-hand side of the board is kept in contact with the nail at all times and that the top left-hand corner of the end of the board is always level with the top edge of the cardboard.

7. Eventually the multiple lines drawn, when extended, will link and the resulting shape will constitute a half-parabola.

8. A 2-foot piece of the half-parabola drawn is cut out carefully and is used as an accurate guide for drawing a full parabola later.

HOW TO DRAW A PARABOLA

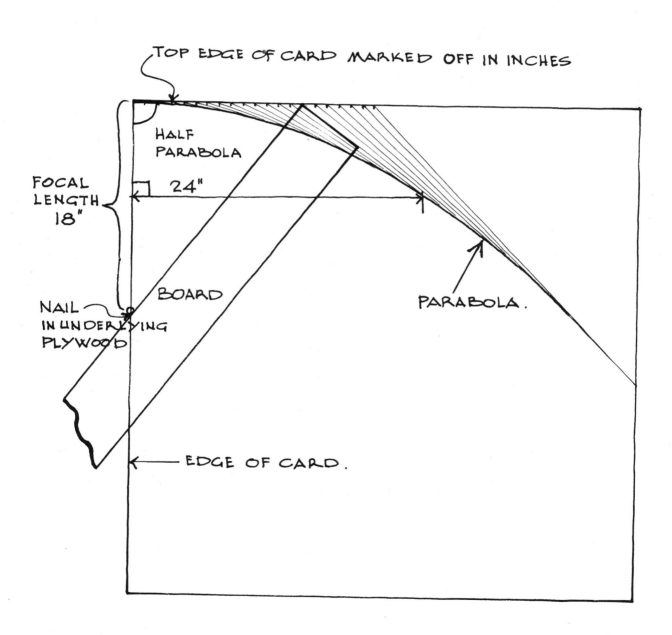

Cutting the Parabolic Frame

Take a sheet of ¾-inch plywood and rip two pieces 48 by 8¼ inches. Mark the middle point of the boards; i.e. 24 inches. Place the half-parabolic shape on the middle line of a board and trace the half-parabola shape. Complete the total parabola by flipping over the half-parabola and tracing the outline. Using the same method, trace out the other parabola. A jigsaw or bandsaw can then be used to cut out the parabolic shapes.

If you put both parabolic shapes side by side you can see how accurate your work was. Accurate or not, the two shapes can be clamped together and sanded so that they are identical.

Reflector Backing

Next rip a strip of hardboard ⅛ inch thick and 13⅝ inches wide. Cut the hardboard to a length of 51¼ inches. Put beads of glue on the parabolic shapes and, using plasterboard nails, carefully fix the hardboard (smooth side out) onto the curved edges of the parabolic shapes placed 12⅛ inches apart. The hard stage of the project is completed; don't give up!

Cut five plywood ribs out of ¾-inch plywood, 13⅝ inches long by 1½ inches wide. These are then glued and nailed onto the back of the hardboard to help support and strengthen it. Place one rib at either end of the parabolas and one at the middle. The remaining two should be placed at equal intervals between the ribs already attached.

Rotisserie Support

The two rotisserie supports are made from 20-by-1½-by-¾-inch plywood pieces. Drill an ⅛-inch hole in each of the 1½-inch-wide sides of each piece of plywood, 2 inches from the ends of the supports. Use a jigsaw to cut a slot from one edge of one support to one of the holes drilled. Nail and glue the rotisserie supports onto one side of each parabola, making sure they are precisely in the middle of the parabolas. Ensure that the slot faces upwards, and both the hole and slot are 18 inches from the inner surface of the hardboard.

Base and Rear Support Struts

Now for the base and rear support strut strips, which are fabricated from ¾-inch plywood. Cut to length two pieces 4 feet long, two pieces 13⅝ inches long, and one piece 4 feet 6 inches long, each piece 2 inches wide. Place the two 4-foot lengths on a bench parallel to each other and 9⅝ inches apart. Nail and glue the short pieces of plywood to join the ends of the two parallel pieces. Drill ¼-inch-diameter holes in all but the last foot of the 4-foot-6-inch rear support strut at 1-inch intervals.

Attach three 1½-inch tight-pinned butt hinges, two to the front of the stand and one to the middle of the rear of the stand, as shown in the illustrations. Place a nail in the middle of the top rib

Hot Dog Cooker

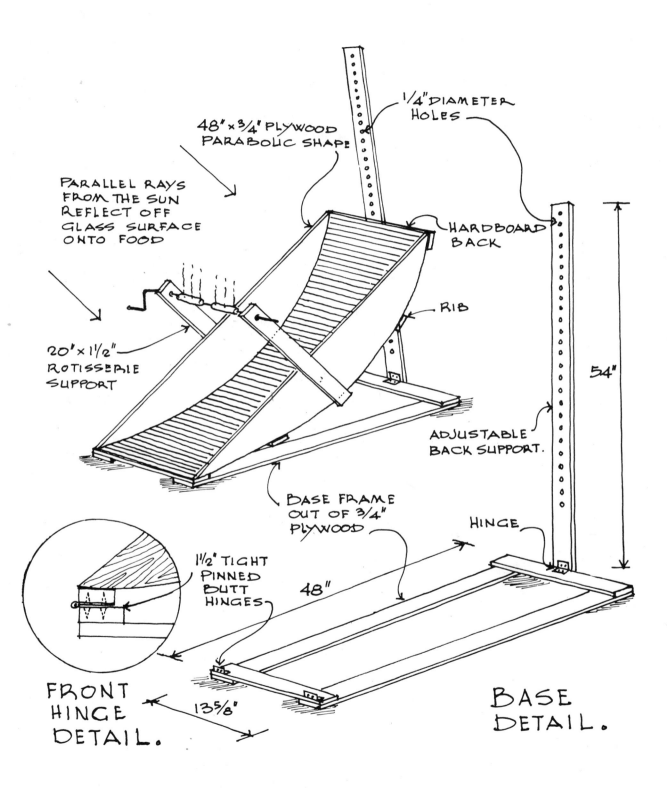

of the parabola. The parabolic shell may now be attached to the base by screwing the two hinges to the shell.

Cutting Mirrors for the Reflector

Now the fun of cutting mirrors! One-foot-square mirror tiles are ideal as a source of mirrors for the cooker and may be obtained from almost any hardware store. Prices vary greatly, so shop around.

Using a felt-tipped pen, mark 1-inch intervals along one side of each of five tiles.

Obtain a good-quality glass cutter (e.g. Diamantor). The mirrors should be placed on a newspaper and the glass cutter's cutting tip dipped in oil prior to cutting. Place one mirror on top of another, leaving a 1-inch strip showing on the lower mirror. Holding the upper mirror firmly and squarely in place, scribe a *continuous* mark with the glass cutter held *vertically*. The cut should be started about $\frac{1}{8}$ inch from the edge of the mirror to avoid damaging the cutting edge. The newspaper will protect the cutter as it comes off the tile. Place a small nail underneath the end of the completed scribe and push down quickly and evenly on either side of the mark. Mirror number one, we hope! Proceed until you have cut at least fifty mirrors.

Mounting the Mirror Strips

Take a square and mark off a series of lines, about 2 inches apart, on the inside of the cooker. These lines will guide you in the placement of the mirrors and will help to keep them square to the edges of the cooker.

Run three beads of clear silicone down the inner hardboard shell; one bead down the middle, the other two an inch from either side. Gently press each of the one-inch strip mirrors into place, taking care to keep them square.

Installing the Spit

Place a thin 15-inch-long stainless steel rod or a straightened coat hanger into the hole and slot of the rotisserie supports, and *voilà!* you are ready for *le hot dog*.

Painting or varnishing completes the project.

Super Hot Dog Cooker

The super hot dog cooker operates in exactly the same manner as its standard counterpart. It is larger in width and has a greater ability to concentrate the sun's energy, allowing you to cook more food faster. Its increased size also makes the project somewhat more expensive to construct.

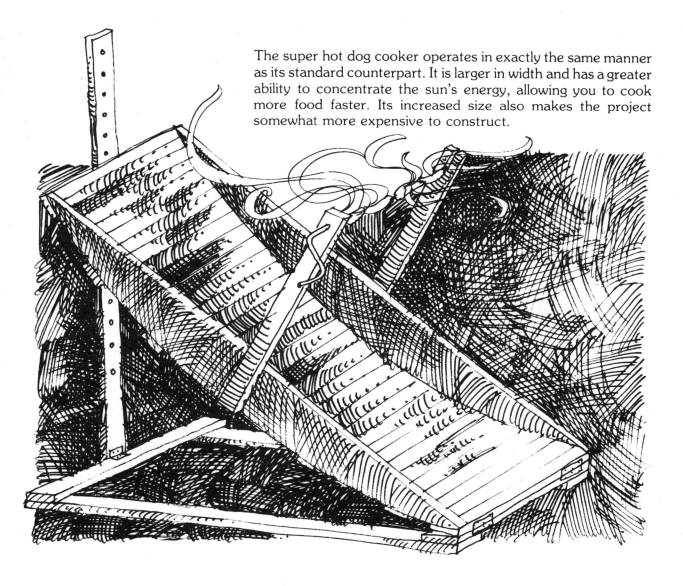

Materials Required

To build this cooker follow the instructions for the standard hot dog cooker with the exception of the modifications outlined below.

Again, ¾-inch plywood is suggested as the basic material for most of the construction. Rip two pieces of the plywood, 5 feet long and 10 inches wide.

1 4-by-8-foot sheet of ¾-inch plywood

1 25⅝-by-72-inch piece of ⅛ inch hardboard or thin plywood wallboard. This is to be trimmed to length later.

65 1-by-24-inch strip mirrors (cut from scrap mirror pieces)

3 2-inch loose-pinned butt hinges with screws

1 30-inch steel welding rod

1 large tube of clear silicone sealant

½ pound of 1¼-inch ring nails

waterproof glue

paint

glass cutter

Reflector Frame

Using the method described previously, mark out a parabola, in this case having a focal point of 24 inches and total parabola length of 60 inches. Using this shape, mark out the two parabolas necessary for your construction and cut them out. Seven equally spaced ribs 25⅝ inches long and 1½ inches wide are used to support a 25⅝-inch-wide band of ⅛-inch hardboard, which is trimmed to length once nailed to the parabolic shapes.

Rotisserie Supports and Base

The rotisserie supports are 2 inches wide and 26 inches long. The back support is 3 inches wide and 5 feet in length. The base is made from two pieces of ¾-inch plywood 54 inches long by 4 inches wide. The crosspieces on the base are 25⅝ inches wide.

Reflective Surface

The super hot dog cooker we made had 24-by-1-inch mirrors mounted in it. The mirrors were cut from scrap pieces obtained from cooperative glass shops. Reflective self-adhesive aluminized Mylar obtained from Edmund Scientific or a graphic arts supply shop may be substituted for the mirrors; but it is expensive. Aluminized paper, used commonly as a reflective vapor barrier, can also be used but is not really as satisfactory as it tends to wrinkle and tarnish, particularly when fat splatters on it during cooking. Mirrors are preferred as they are easy to clean and are very long-lasting.

super Hot Dog Cooker

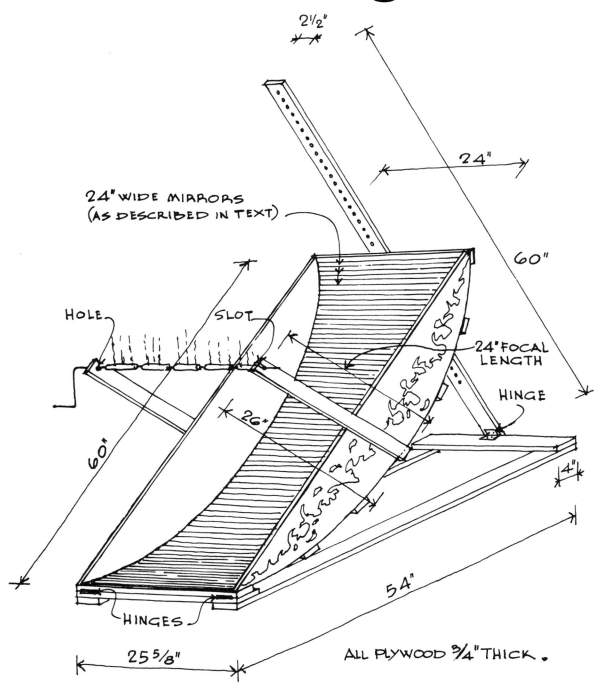

CONSTRUCTION DETAIL.

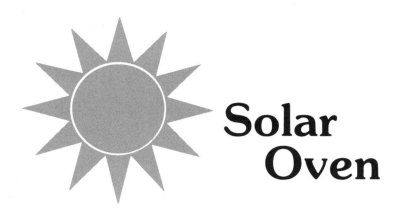

Solar Oven

This solar oven consists of an inner box of stainless steel and an outer box of galvanized iron, separated by 1 inch of industrial fiberglass that helps to retain heat. On the front of the oven is a removable double-glazed "window" that allows light to enter the oven and that traps heat radiating from the blackened interior. Reflective panels on the front of the oven concentrate the sunlight entering the oven. During sunny weather, temperatures of 400°F. are easily achieved.

A stand supports the oven at a convenient height. The oven is attached to the stand by two hinges at the rear of the oven to

allow it to be tilted so that the direct rays from the sun, entering the window of the oven, are always perpendicular to the surface of the glass. Wedges under the oven can be moved in and out to accomplish changes in the inclination of the base of the oven.

The oven that was originally constructed was designed around a secondhand microwave oven door. A window for the oven door could equally well be made using tempered glass from a conventional oven and building a frame around it. The assistance of a skilled sheet-metal worker might well be advised unless you have considerable experience working with sheet metal.

Materials Required

15 square feet of self-adhesive aluminized Mylar (at least 2 feet in one dimension)

10 square feet of light-gauge stainless steel (at least 2 feet in one dimension)

25 square feet of light-gauge galvanized sheet metal (at least 2 feet in one dimension)

1 24-by-30-inch piece of ¾-inch plywood

20 feet of 1½-by-1½-inch softwood

1 recycled microwave oven door or equivalent

4 dozen ⅜-inch sheet metal screws

1 tube of clear silicone sealant

2 2-inch loose-pinned butt hinges

2 metal handles for oven door

2 wooden wedges (see illustration and text)

industrial fiberglass sufficient to insulate structure (This insulation is usually available from heating equipment suppliers.)

rivets

paint

Internal Box

The internal box should be made as follows. The various pieces of light-gauge stainless steel are cut to the dimensions shown in the exploded view of the oven. Each of the protruding flanges is made 1 inch wide and should be bent at 90° except where noted otherwise. All screws should be fixed in position by screwing them from the inside of the box outwards. Using ⅜-inch number 6 sheet metal screws, attach the top to the two sides. Then firmly secure the back in place. Next slide the base into position and screw it to the remainder of the structure already constructed. Finally, attach the front to the internal box.

Solar Oven

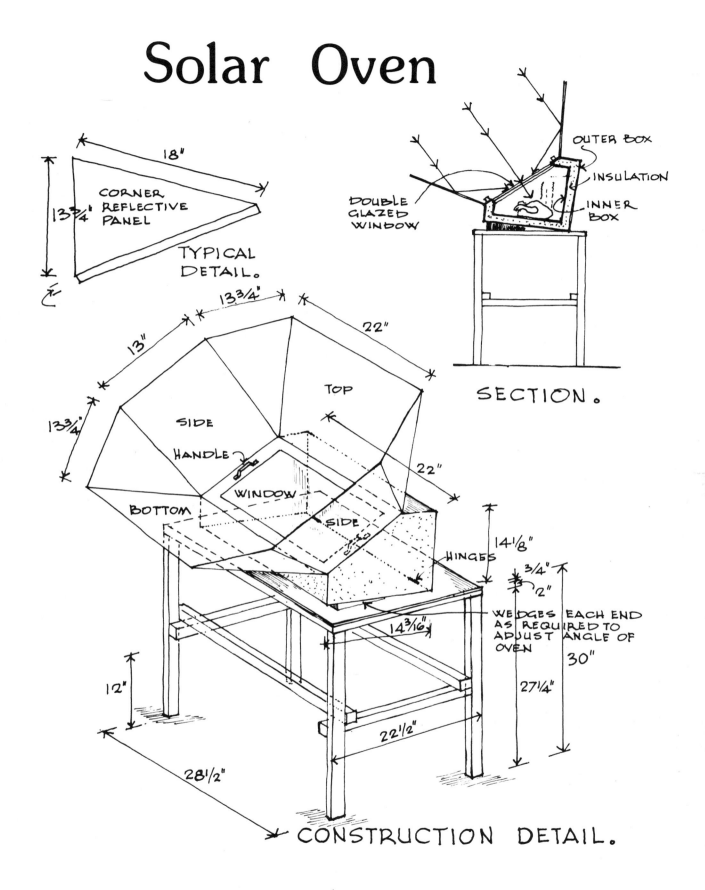

Solar Oven

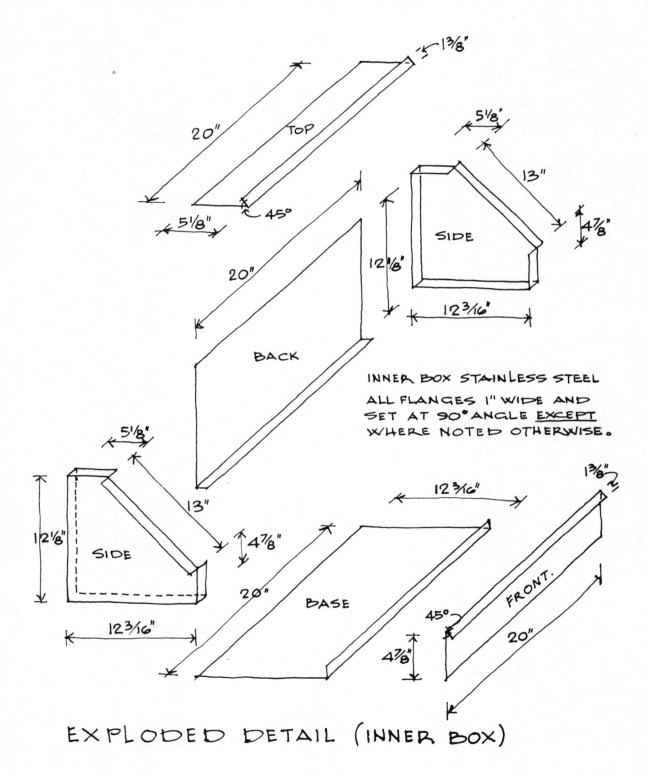

EXPLODED DETAIL (INNER BOX)

Solar Oven

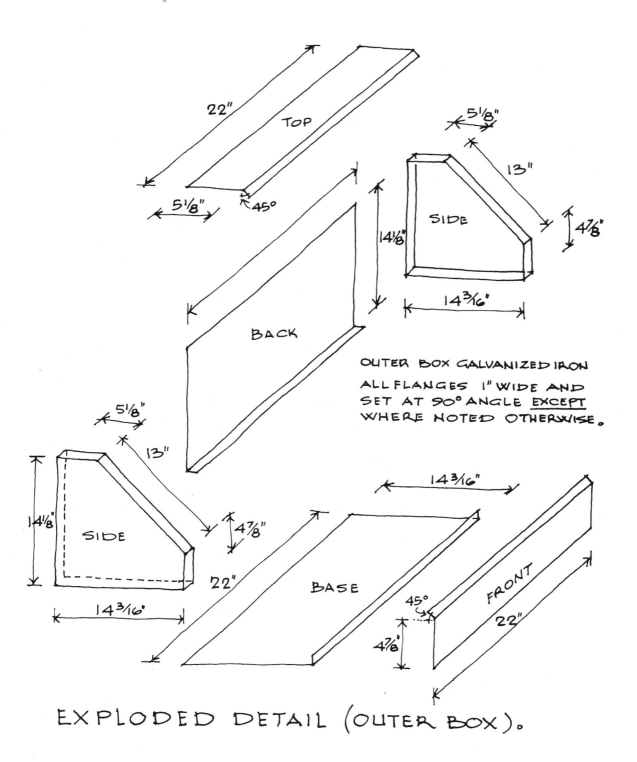

EXPLODED DETAIL (OUTER BOX).

External Box The external box is made of galvanized sheet metal. Follow the illustrations and the order of assembly outlined for the internal box. All screws are screwed from the outside into the interior. Before the front is fixed in place, slide the internal box inside the box you are fabricating. At the same time stuff the cavity with industrial fiberglass. Use screws to secure the two boxes together where they overlap. Cut small metal plates to cover each gap remaining at the corners, then braze into position.

Oven Door Side flanges are screwed to the microwave oven front and handles are placed on either side of it. The front of your oven will have to be made according to the materials available to you. Bear in mind that the glazing must be heat-resistant. Whatever window arrangement you fabricate, make side flanges up to 1 inch in width and the length of the sides of your oven window frame, to hold the frame in place.

Reflective Panels The reflective panels are made of light-gauge sheet metal. The top and bottom reflective panels are cut 18 by 22 inches and the two sidepieces are cut 18 by 13 inches. The four corner reflective panels are cut in a regular triangular shape; two sides 18 inches long and base 13¾ inches. The 18-inch sides have 1-inch extensions on them (see illustration). All the reflective panels are riveted together, and then hinges or sheet metal pieces are used to join the reflector configuration to the front edge of the oven. The interior surface of the panels is lined with self-adhesive reflective Mylar.

Stand The stand is made as follows: cut the top from ¾-inch plywood, 24 by 30 inches. Then cut two pieces of ¾-inch plywood, 2 inches wide and 24 inches long, and two pieces 28½ by 2 inches. Nail the pieces together to complete the top of the stand (as illustrated). Cut four legs 30 inches long out of 1½-by-1½-inch softwood. Glue and screw these legs under the corners of the stand top. Cut two stand crossmembers 22½ inches long; again from 1½-by-1½-inch lumber. Glue and screw into place 12 inches from the bottom of the legs, as shown in the diagram. Then attach two more crossmembers, measuring 28½ by 1½ by 1½ inches, as shown.

Finally, attach hinges to the back of the oven connecting the oven to the stand. Cut two wooden wedges with which to adjust the angle of the oven. The farther the wedges are pushed under the oven the greater the inclination becomes.

If high cooking temperatures are desired, the whole oven may be covered with a removable insulating cover to minimize heat losses.

Bon appétit; the menu is as varied as your creative imagination!

Chicken Cooker

This solar chicken cooker is capable of cooking a 3-pound chicken in about an hour and a half under ideal conditions. Light falling on the internal surface of the parabolic mirrored surface is reflected onto a focal point. The chicken protrudes outwards from the focal point and therefore does not receive the full concentration of the sun's rays. A rotisserie slowly rotates the chicken as it cooks.

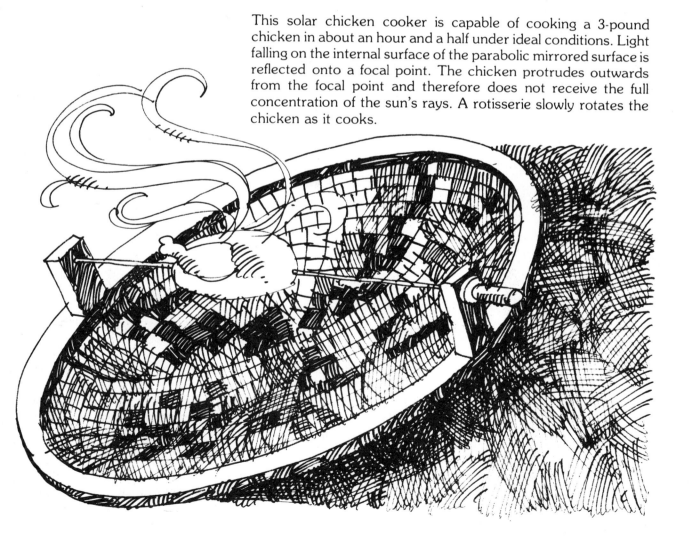

The sun's rays can be focused by adjusting the position of the parabolic shape by first rotating the support stand and then moving the parabola up and down using the chain adjustment.

Materials Required

2 4-by-8-foot sheets of ½-inch plywood

1 4-by-8-foot sheet of ¾-inch plywood

1 4-by-8-foot sheet of ⅛-inch hardboard

25 pounds of plaster of paris for parabolic shell (or fiberglass resin, matt and catalyst)

25 pounds of plasterboard filler for dome mold construction

450 2-by-2-inch mirrors cut from scrap mirror pieces

15 square feet of ½-inch welded steel mesh

1 vehicle brake-drum assembly

3 feet of 2-by-4-inch softwood

1 ⅜-inch-diameter steel rod 52 inches long

1 5-foot heavy-duty metal chain

1 ½-inch-diameter 5-inch bolt

1 rotisserie motor; battery or 110v

1 caulking tube of clear silicone sealant

4 heavy-duty castors

1 1-by-1-inch galvanized plate

1 pound of 1¼-inch galvanized finishing nails

waterproof resin glue

½-inch staples

petroleum jelly

paint

Parabolic Reflector

First a mold must be made so that a reflector shell may be accurately formed. Mark and cut out fifteen half-parabolas, of ½-inch plywood, each half-parabola being 2 feet across and having a focal length of 18 inches. (Follow the details of parabola layout outlined under the construction of the hot dog cooker.)

Draw a circle on the ground, having a diameter of 4 feet. Symmetrically arrange the fifteen half-parabolas in a circular pattern to produce a dome shape. Use individually cut pie-

Chicken Cooker

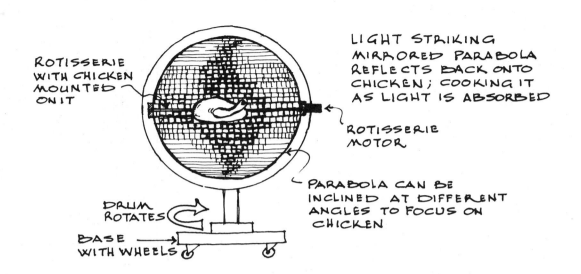

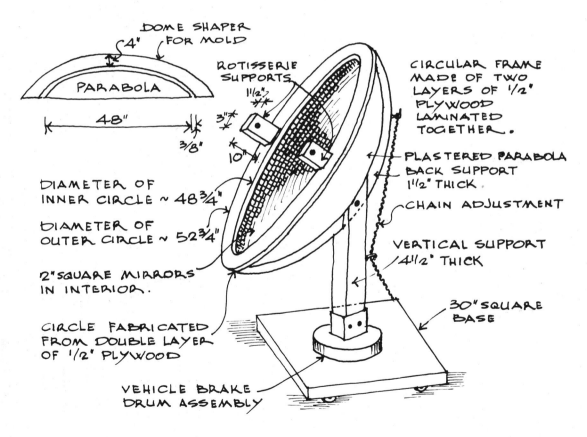

CONSTRUCTION DETAIL.

shaped sections of ⅛-inch hardboard to cover the shape on the floor.

Cover the dome-shaped mold with several layers of plaster; plasterboard plaster does a good job. Fabricate a piece of wood that will be used to accurately finish the mold. Using your original parabolic shape (focal point 18 inches, width 4 feet), produce this plywood shape from ¾-inch plywood. Its inner parabolic surface should be ⅜ inch away from the outer surface of the original parabola used to produce the dome (see illustration). This "dome-shaper" board is then carefully rotated over a thin layer of wet plaster to obtain an accurate mold. The dome-shaped mold is smoothed off when dry and is then painted with a hard, high-sheen enamel paint.

Front Frame

Next comes the fabrication of the front frame of the cooker. Draw two concentric circles on ½-inch plywood having radii of 24⅜ inches and 28⅜ inches. It will be impossible to cut a whole circle out of one sheet of plywood. The shape has to be made up of six segments of ½-inch plywood. The inner partial circumference of each piece of plywood is 51 inches. Three of these shapes are glued and nailed to three identical shapes underlying them. Make sure the joints on the underlying circular shapes are in between the joints of the top circle.

Back Support

Take the plywood dome-shaper and cut another identical piece of ¾-inch plywood. Laminate the two pieces together with glue and nails, and attach with screws to the circular shape that you have already made. Glue and screw 1½-by-3½-by-4-inch blocks to either side of the back support where it joins the circular shape to hold it rigidly.

Forming and Plastering the Reflector Shell

Staple ½-inch welded iron mesh all over the interior of the shape you have now completed. Check frequently to see that the shell being formed out of the wire will clear the mold when you place the shell over the mold.

When all stapling is completed, place the shell over the mold, which has first had petroleum jelly applied all over its surface. Plaster is then forced through the wire until a thickness of some ½ inch is achieved. Allow to dry thoroughly and then remove carefully. Use solvent to remove any petroleum jelly remaining on the parabolic shell. Replaster and sand to obtain a perfect finish inside and outside the shell. (Fiberglass could be used instead of plaster and would probably give a more permanent finish.)

Installing Mirrors

Cut about 450 mirrors, 2 by 2 inches, from scrap pieces obtainable from glass shops. Use clear silicone to secure these mirrors to the shell. The first line of mirrors should be placed along any straight line passing from one side of the shell, through the center, and on to the opposite side of the shell. Keep putting mirrors into the shell until its surface is totally mirrored. Clean the mirrors with a razor blade twenty-four hours after they are completely set in place. Traces of silicone can be removed by rubbing the mirrors with steel wool followed by a cloth soaked in vinegar.

Rotisserie Support

Rotisserie supports are made from 3½-by-1½-by-10-inch wood. Drill ⅜-inch holes through the 3½-inch face of each of these supports. The holes should be centrally located 1 inch from the end of each support. Mount the rotisserie supports as illustrated; secure firmly to the circular frame. Screw a small galvanized plate over the outer end of one of the holes in one rotisserie support to keep the rotisserie rod from coming out of the support. The exact length of the ⅜-inch-diameter rod is measured according to the rotisserie motor you can obtain to rotate it.

Vertical Reflector Support

The vertical support is made from multiple layers of ¾-inch plywood which are glued and nailed together. First cut two pieces 24 inches long and 6 inches wide and laminate these pieces together. Cut the end of the plywood, as illustrated, so that one side is 24 inches long and the other 19 inches. Four further pieces of ¾-inch plywood are cut 6 inches wide, 30 inches long on one side, 25 inches on the other. Glue and nail two of these pieces on one side of the support pieces already laminated, and two on the other side. A groove will then be formed to accommodate the back support of the mirrored parabola shell.

Drill a ½-inch hole in the back support 2 inches from the back edge of the support's midpoint. Also drill matching ½-inch holes 1 inch away from the midpoint of the top sloping edge of the support stand. A 5-inch-long ½-inch bolt will later hold the vertical support and back support together.

Base

The base is made from two 30-by-30-by-¾-inch plywood pieces glued and nailed together. Four castors are screwed to the base; these should be placed close to the corners.

The brake drum assembly used to allow the stand to rotate will have to be custom-fitted to the base and stand according to the size of the unit available.

A chain is secured to the back support by means of a screw-threaded hook. Another screw-threaded hook is screwed into the support stand. The collector is now fully adjustable.

Paint or stain your chicken cooker as desired. Your cooker is ready for its first cookout. Have fun!

Parabolic Concentrating Cooker

This parabolic concentrating cooker is basically a solar furnace. It can easily be used for cooking or other experimental purposes. The mirrored parabolic surface reflects the light, bringing it to a fairly sharp focus, depending upon the size of the mirrors used and the accuracy of construction. The whole mirrored parabolic shell, mounted on a pipe which pivots on a stand, can be readily adjusted to focus the sun. A rotisserie can be mounted on the furnace if you wish to use it for cooking.

Materials Required

1 4-by-8-foot sheet of ¾-inch plywood

2 4-by-8-foot sheets of ½-inch plywood

1 4-by-8-foot sheet of ⅛-inch hardboard

450 2-by-2-inch mirrors (cut from scrap mirror pieces)

80 feet of 1-by-1-inch softwood

2 feet of 1½-by-1½-inch softwood

1 56-inch-long 1-inch-diameter steel pipe

10 pounds of plasterboard filler

4 6-inch steel brackets

1 pound of 1½-inch galvanized nails

½ pound of 1¼-inch galvanized nails

8 1-inch pipe fasteners

1 1-inch threaded hook

1 caulking tube of clear silicone sealant

waterproof glue

paint

Parabolic Shell Frame

A 4-by-4-foot piece of ¾-inch plywood is cut out to form the base for the parabolic shapes upon which the mirrors will be mounted. Draw a 4-foot-diameter circle on the sheet of plywood.

Rip out 20 pieces of ½-inch plywood, 2 feet long and 10 inches wide. Using a parabolic shape with a focal point of 18 inches (following the instructions for the hot dog cooker), mark out half-parabola-shaped ribs on the 20 pieces of plywood. Glue and nail 1-by-1-inch strips onto each of the parabolic ribs along one side of the long straight edge of each rib. Then glue and nail all these ribs to the base so that each one radiates from the center, 7½ inches on center, around the circumference. Trim the narrow ends of the ribs as necessary to fit them all on the base.

Inner Shell

Hardboard, ⅛ inch thick, is used to cover over the parabolic ribs and form an inner shell. Push the hardboard down onto two ribs at a time and use a pencil to mark the shape of a section of the hardboard by running the pencil line along the outside of the top edge of the two ribs in question. Using a bandsaw or jigsaw, cut the section of hardboard ¼ inch inside the lines marked. Glue and nail the hardboard in place. Repeat the process until all the shell is covered with hardboard. Glue and nail a 10-inch-wide hardboard trim around the vertical perimeter of the cooker shell. Trim any excess plywood from the base so that 2 inches protrudes beyond the vertical trim just completed.

Installing Pivot Shaft and Finishing the Shell

Using pipe fasteners secure a 56-inch-long 1-inch-diameter steel pipe to the base of the collector. Put a thin layer of plaster over the inside of the shell and rotate a complete parabola (4 feet wide, 18 inches focal point) over the plaster surface until it is smooth; sand when dry. Using silicone, mount 450 2-by-2-inch mirrors onto the completed shell of the furnace. Follow the same procedure for mounting and cleaning the mirrors as outlined for the previous project.

If you want to mount a rotisserie, then the holes supporting the rotisserie must be arranged so that they are 18 inches above the center of the reflecting surface of the collector.

Stand

The stand is constructed from plywood. The following pieces are cut: two base side struts, ¾ by 6 by 48 inches; two base cross supports, ¾ by 6 by 55 inches; and two vertical supports, made from four pieces of ½-inch plywood, 6 inches wide and 26 inches high, having notches 1 inch wide and 3 inches deep cut into their

Parabolic Concentrating Cooker

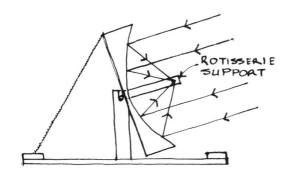

MIRRORED SURFACE REFLECTS LIGHT ONTO FOOD 18" FROM SURFACE.

PARABOLIC RIBS COVERED WITH HARDBOARD WITH MIRRORS ON SURFACE OF 48" DIAMETER CIRCLE (CIRCUMFERENCE 151").

TYPICAL RIB

½" PLYWOOD STRIPS BOTH SIDES

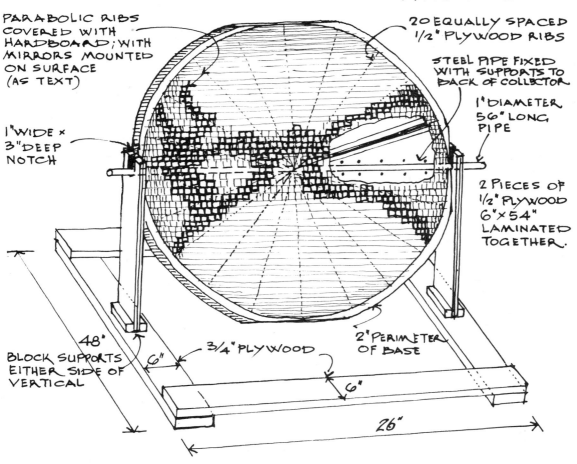

PARABOLIC RIBS COVERED WITH HARDBOARD; WITH MIRRORS MOUNTED ON SURFACE (AS TEXT)

20 EQUALLY SPACED ½" PLYWOOD RIBS

STEEL PIPE FIXED WITH SUPPORTS TO BACK OF COLLECTOR

1" DIAMETER 56" LONG PIPE

1" WIDE × 3" DEEP NOTCH

2 PIECES OF ½" PLYWOOD 6" × 54" LAMINATED TOGETHER.

48"

BLOCK SUPPORTS EITHER SIDE OF VERTICAL

6"

¾" PLYWOOD

2" PERIMETER OF BASE

6"

26"

CONSTRUCTION DETAIL.

top ends. Assemble the base, gluing and nailing the pieces together as shown in the illustration.

Cut four 6-by-1½-by-1½-inch block supports; glue and screw these on either side of the verticals to give them stability. Steel brackets may also be added to attain further strength and rigidity if desired. Use hooks to attach a chain to the middle of one of the base cross supports, and to the base of the collector structure. Adjustments in the inclination of the collector are then easily made by altering the length of the chain.

Paint the whole cooker-furnace according to your preference and you will then be ready for cooking or experimenting. Try putting a piece of dark colored wood at the focal point; surprising results!

Water and Air Heaters

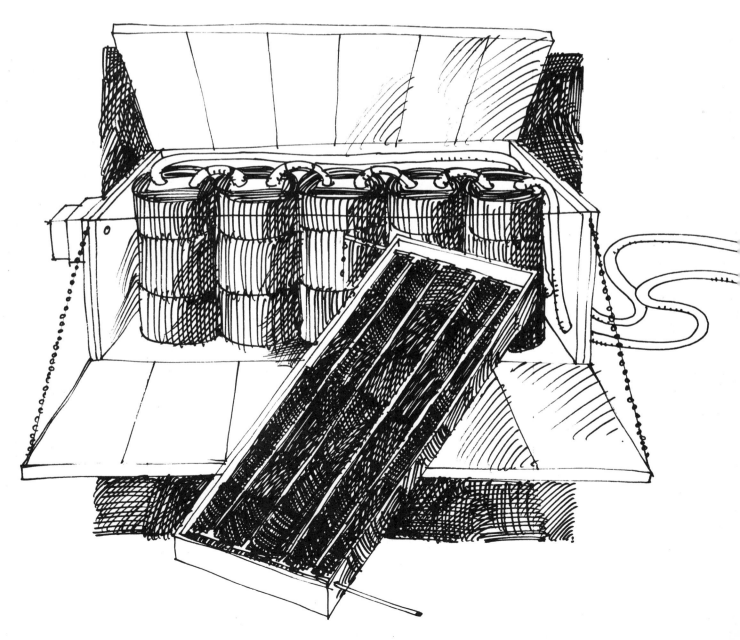

Water Preheating Panel

This solar water preheater panel is an example of a collector panel that could be used in conjunction with two other panels of the same size to supply much of the hot water required by a family of four.

It is basically a double-glazed unit that has in its interior copper pipes soldered to standard galvanized roofing. (There is some question of the long-term durability of a soldered galvanized-copper joint; thus far we have had no problems.) The blackened surface of the galvanized iron heats up and transfers its heat to the copper pipes attached to the sheet. Water flowing through the pipes carries away the solar energy collected. Cold water enters the collector via the bottom header and flows evenly through each of the risers, exiting, after heating, from the top header. Providing water flow rates are not too high, the collector can easily produce water at 120°–140°F.

The double glazing and insulation in the collector help keep heat losses to a minimum. These losses are further minimized by using silicone to mount the collector plate in position. The use of silicone mounting also allows for free expansion and contraction of the collector plate.

Materials Required

1 4-by-8-foot sheet of ¾-inch plywood

1 2-by-8-foot piece of 3/4 inch plywood

24 feet of 16-inch-wide R–12 (4 inch) fiberglass batt insulation

1 8-foot sheet of standard galvanized roofing

1 96-by-38½-inch sheet of Mylar or Tedlar

1 96-by-40-inch sheet of Kalwall premium

32 feet of ½-inch I.D. hard copper tubing

8 feet of ¾-inch I.D. hard copper tubing

8 ¾-by-¾-by½-inch copper tees

2 ¾-inch copper caps

24 feet of 7½-inch-wide light-gauge galvanized sheet metal (to flash collector-box edges)

1 4-by-8-foot sheet of light-gauge galvanized sheet metal (if required)

24 feet of ¾-by-¾-inch vinyl coping

24 feet of ¾-by-¾-inch redwood to support inner glazing

½ pound of 1½-inch galvanized finishing nails

24 ⅜-inch sheet metal screws

1 can flat black heat-resistant paint

1 tube of clear silicone sealant

1 small can of contact cement

waterproof glue

wood preservative

Collector Plate Supporters   Cut three collector plate supporters ¾ by 3½ by 38½ inches and install them in the collector box: one at the center and the other two 12 inches from either end of the box. Firmly glue and screw the supporting strips in place. Mylar supporting strips should also be secured on the inside of the collector box ½ inch from the top. Two of the strips are 94½ inches long, the other two 37 inches. Paint all wooden surfaces completed with two coats of wood preservative.

Collector Plate   The actual collector plate, 96 by 34½ inches (i.e. standard galvanized roofing), is reduced in length to 89 inches by cutting with tin snips. Four ½-inch-I.D. straight copper tubing pieces are cut, each 87 inches in length. Thoroughly clean the surface of the galvanized sheet and copper with steel wool before soldering the copper tubes in place. When soldering use a paste-type flux and 50:50 solder.

Soldering the Tubes   The best method of accomplishing the soldering is to clamp the copper tubes in place using a steel tube, placed across the collector, which is clamped on either side. A short length of each copper tube is then soldered and allowed to cool, after which the

Water Pre-heating Panel

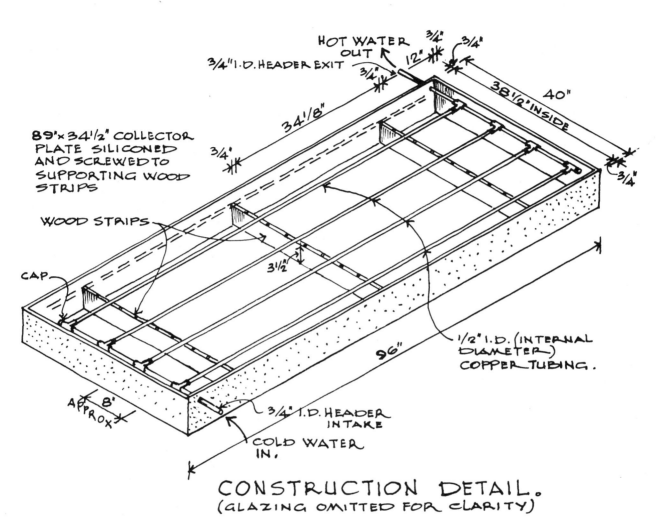

CONSTRUCTION DETAIL.
(GLAZING OMITTED FOR CLARITY)

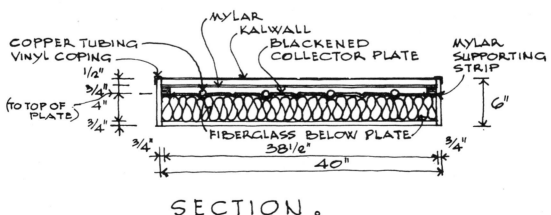

SECTION.

steel tube and clamps are moved further down the collector and the process is repated. A 2-inch section at both ends of each copper tube should be left unsoldered. Bend these short sections slightly upwards to accommodate the T's which may then be slipped over them.

Measure the exact distances between the ends of each copper tube and cut ¾-inch-I.D. copper sections to fit in the header configuration. Solder the first T at the end of one pipe and then proceed to build up the whole header configuration. Do not put all the T's on at first otherwise you will not be able to fit the other straight sections in between. Put a 2-inch extension on the end of each header, as illustrated, and cap them.

Drill 1-inch holes to allow the inlet and outlet pipes to enter and leave the collector; these holes should be drilled once you know precisely where they are to go in relation to the rest of the header arrangements.

Installing Insulation and Collector Plate

Using 4-inch industrial fiberglass, pack the bottom of the collector and the sides and ends. Silicone the top of each plate supporter; then, using sheet-metal screws, secure the plate to the supporters. Twenty-four hours later remove all screws except the middle ones so that the plate can move without buckling during expansion and contraction.

Then solder the copper inlet and outlet tubes in place. Measure these tubes so that they extend at least two inches outside the box.

Glazing

Now for the glazing. Silicone and staple a Mylar sheet to the supporting strips already in place. Make sure that this sheet is stretched as tightly as possible to prevent later sagging. To complete the glazing, use Kalwall premium siliconed in position and futher secured by heavy-duty vinyl coping.

To complete the project the collector should be caulked, flashed with galvanized steel, and then thoroughly painted with a premium-quality paint.

Demonstration Hot Water Heater

The original idea for the construction of a demonstration hot water heater, using a refrigerator heat exchanger and windshield washer pump, came from Tom Walton, physics lecturer at Cariboo College, Kamloops, Canada.

This hot water heater is a unit that is ideal for use in science fairs or as a general unit for demonstrating many basic principles involved in collector design. The basic design consists of heat exchanger coils from a refrigerator that form a collector when painted with a heat-resistant flat black paint. Removable double-glazing is placed over the collector. Insulation behind and at the sides of the collector reduces heat losses. Adjustable pipe supports are used to alter the inclination of the collector. The windshield washer pump is used to circulate the water through the collector coils. The flow rate can be changed by the use of a variable power source.

It is possible to test many variables because of the flexible nature of the basic design. Specific construction details are not possible as the heat exchanger available will vary in dimensions; however, generalizations concerning the design may be made.

Materials Required

1 sheet of 4-by-8-foot pressure-treated ¾-inch plywood

1 recycled heat exchanger from a refrigerator

Glazing: Outer, U.V.-inhibited fiberglass; Inner, Tedlar

12 feet of 2-by-4-inch softwood

1 4-foot 1¼-inch O.D. steel tube

1 4-foot-15/16-inch I.D. steel tube plus 2 adjustment screws (see illustration)

1 windshield washer pump

1 12-v d.c. variable power source (110-v a.c. supply)

4 2-inch hinges plus screws

1 4-by-2-foot (approximately) piece of 4-inch (R–12) fiberglass insulation

10 feet of self-adhesive foam weatherstripping

6 feet of rubber tubing (sized to fit heat exchanger tubes)

1 small can of flat black heat-resistant paint

½ pound of 1½-inch galvanized nails

2 dozen 2½-inch galvanized nails

waterproof glue   paint

Demonstration Hot Water Heater

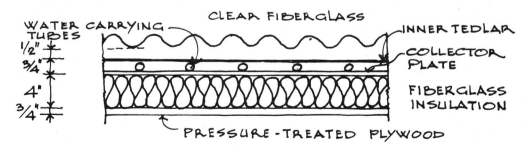

CONSTRUCTION DETAIL.

TYPICAL SECTION

Collector Box and Plate

Pressure-treated plywood is an ideal material from which to build the collector box; ¾-inch material is recommended. The collector plate is silicone mounted (for thermal insulation) on three supporter struts which are placed one at the center and two close to either end of the collector. These supporters should be 4 inches deep, ¾ inch thick, and the width of the collector. Industrial fiberglass (4-inch) is packed behind the collector plate and 2 inches to either side of the plate.

Glazing

A gap of approximately ¾ inch is left between the collector plate and a removable frame holding du Pont Tedlar. The outer glazing is ultraviolet-inhibited corrugated fiberglass which is mounted on a hinged frame that is weather stripped and held in position by hooks and eyes.

Supporting Base

The supporting base consists of a 2-by-4-inch framework which is hinged at the front and rear, the latter hinges being welded to tubular inclination adjusters. The two pipes comprising the adjusters fit inside one another, which allows the angle of the face of the collector to be easily adjusted. When the desired angle is obtained, a screw adjustment threaded into the outer pipe secures the inner pipe. Hinges welded to the top of the adjusters and screwed into the rear of the collector complete the stand's structure.

Pump and Storage Tank

The pump and storage tank are placed in an insulated box. Water is pumped up the collector's serpentine tubing and back down the side of the collector. Pumping upwards ensures that all tubing is full of water and virtually eliminates the possibility of hot spots caused by air pockets in the tubing. A variable power source, as previously mentioned, may be used to control flow rate.

A variety of experiments can be performed using the collector by varying the glazing, the angle of inclination of the collector, the flow rate through the collector, etc. A great number of different variables may be tested while other factors are kept constant.

Barling's Barrel Bread-Box Heater

A do-it-yourself bread-box heater can be a low-technology solar water heater that is relatively simple to build. At the same time it is a very cost-effective structure that produces large volumes of hot water. The description of this heater does not give specific construction details, but rather gives details of a conceptual model that is going to be built by the author.

This bread-box heater is both a collector and hot water storage area in one. Basically the structure consists of a shoe-box-shaped device that is oriented west to east along its long axis. The well-insulated box (R–20) is double-glazed on both its top and its south-facing side. Sunlight enters the box directly or is reflected off the reflector panels into the box. The light entering the bread box strikes the blackened drums or reflective aluminum box lining. The drums heat up, when the light is absorbed, and transfer heat to the water inside them.

Barrels: 5 used 45-gallon fiberglass resin drums

Sheathing: ½-inch pressure-treated plywood

Reflective material: highly polished used aluminum printing plates

Glazing: Outer, Kalwall premium or tempered glass; Inner, Tedlar

Insulation: 4-inch (R–20) Styrofoam on reflector panels; 6-inch (R–20) fiberglass in walls of bread-box

Copper tubing: 70 feet of ½-inch I.D. soft copper tubing

Collector and Storage

Each of the steel 45-gallon fiberglass resin drums (cost about $2) is 90 percent filled with water and nontoxic propylene glycol antifreeze. The air space at the top of each barrel is to allow for expansion of the water.

Running through each barrel is a ½-inch-I.D. copper tube. In each barrel is a 10-foot coil of the copper to achieve good heat exchange. The water in the coils is under pressure and enters from the house plumbing. The cold water running through the coils is progressively heated and leaves the bread box, where it travels directly into the hot water pipes, or alternatively into a conventional water heater. An air-filled expansion tube should be fitted into the piping to allow for expansion as the water is heated if there is a pressure-reducing valve located where cold water service enters your home.

The coils are removable from each drum because each is soldered to a galvanized sheet which is siliconed and screwed into position in the top of each drum. Each coil has a coating of tar on its exterior to avoid any potential long-term galvanic reaction with the iron of the drum.

If the bread box is not used in the winter, the copper coils can be emptied by blowing compressed air through them, followed by hot air (e.g., from a hair-dryer). The antifreeze would then protect the drums from freeze-up damage. (Check to see that your local code allows the use of antifreeze next to pipes containing potable water).

Throughout spring, summer and fall, the bread-box heater should provide hot water reserve supply (200 gallons) for two cloudy days, for a family of four. The percentage of hot water a system such as this could supply would depend on the climate and many other variables.

Reflector Panels

The reflective panels above and below the bread box reflect light into the bread box during sunny weather. During poor weather and at night, the closed panels reflect heat back into the interior, and the heavy insulation in the whole structure further helps to retain heat.

The opening and closing of the reflector panels could, if desired, be made fully automatic by coupling each panel to an electric motor that would be controlled by a differential thermostat. One probe of the thermostat would be housed in the actual bread box and the other would be externally mounted in a mini-collector. Once the temperature was 10°F. higher in the mini-collector than in the bread box, a time-delay relay would activate the motor and open the reflectors. Conversely, as soon as the temperature in

Barling's Barrel Bread-Box Heater

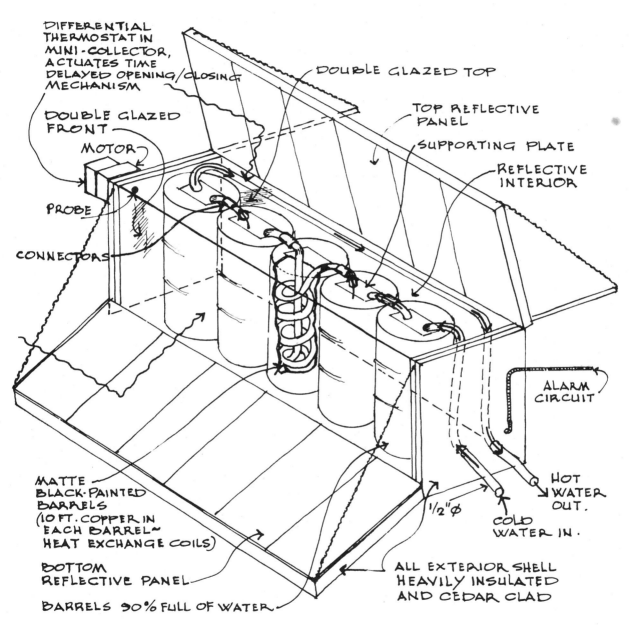

CONSTRUCTION DETAIL.

the mini-collector dropped 10°F. below that of the bread box the reflector panels would automatically close. The mini-collector would have less insulation than the bread box to make the probe in it more responsive to temperature drops, and so the reflectors would be closed fairly quickly, reducing unnecessary energy losses.

The steel drums in the bread-box should last at least 20 years, providing the metal is not continuously exposed to oxygenated water. As long as the drums are not opened, any oxygen that was already dissolved in the water will react with the iron of the drum and then further rusting will be limited. If a leak should occur, the water released would complete a circuit and sound an alarm in the adjacent home.

I believe that this bread box heater has considerable potential as an inexpensive, practical source of domestic hot water.

Solar Swimming Pool Heater

A solar swimming pool heater is typically an unglazed flat-plate collector capable of efficiently collecting solar energy and transferring it to the pool to be heated.

The aim of the system is to raise the temperature of a large volume of water so that on average its temperature is about 10°F. higher than it would be without the system.

Water flows through the collectors at high flow rates and is heated up to 5°F. on each pass. Because the water in the collectors is not much higher in temperature than that of the ambient air temperature, heat losses are relatively low. For this major reason glazing is largely superfluous. Glazing also reflects a high proportion of sunlight when the sun is low in the sky, so that glazed collectors would actually collect less solar energy in the early morning and evening than an unglazed system would.

Materials Required

Plumbing to heater

Schedule-40 plastic piping or schedule-160 (N.B. schedule-40 more durable). Diameter of plastic pipe 1½ or 2 inches, dependent on size of system (see text)

1 vacuum-relief valve

Controls

Manual or automatic (see Panel and Plumbing Layout)

Materials necessary for template and form construction

2 4-by-8-foot sheets of ½-inch plywood

½ 4-by-8-foot sheet of ¾-inch plywood

glue

Materials necessary per panel

1 4-by-8-foot sheet of ½-inch plywood

1 4-by-8-foot sheet of ¼-inch hardboard

25 feet of heavy-duty U-configuration vinyl coping

2 2-by-8-foot sheets of 0.135-inch copper

2 16-foot lengths of ½-inch soft copper tubing (N.B. approximately 2/3 of 50-foot coil)

2 4-foot lengths of 2-inch low-pressure copper drainage pipe (2/3 of 12-foot standard length)

4 CxM threaded copper adapters (thread to fit 9/16-inch hole in header; open end to accommodate ½-inch riser tube)

1 caulking tube of clear silicone sealant

1 quart of Rustoleum flat black paint

contact cement

cold tar

50:50 solder

flux

steel wool

Collector Location and Tilt

The panels of the system should be placed on a south-facing roof or a rack custom-made for the job. The tilt of the panels should be latitude minus 10° to 15° to optimize the collection of energy from the summer sun.

As a very rough rule of thumb the panel area for the particular pool should be about 50 percent of the surface area of the pool, providing roof pitch, orientation, wind factor, etc., are optimal. Systems using a small area of panels are of little use because they are not *effective* when they are needed most, i.e., at the beginning and end of the heating season.

Collector Panel

Copper was chosen as the recommended collector material for both the tubes and the flat plates because of its reliability and compatibility with chlorinated water. Aluminum is totally unsuitable for this particular use, as it corrodes far too quickly. A combination of copper tubes and aluminum flat plates produces

Pool Heater

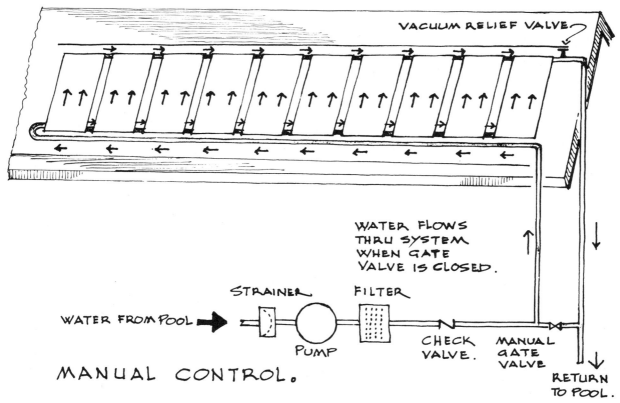

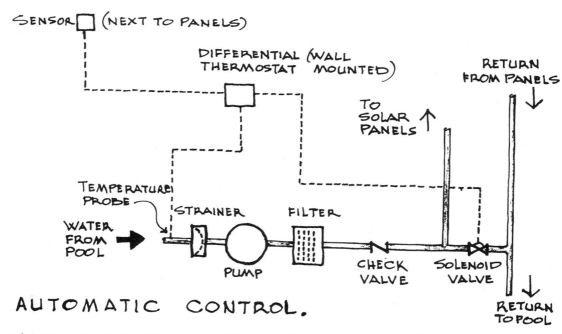

PANEL AND PLUMBING LAYOUT

Pool Heater

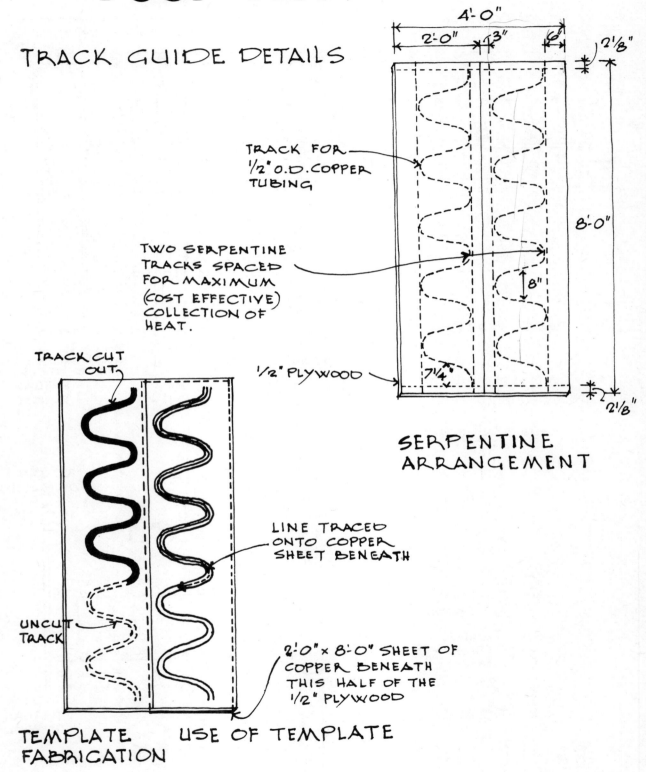

Pool Heater

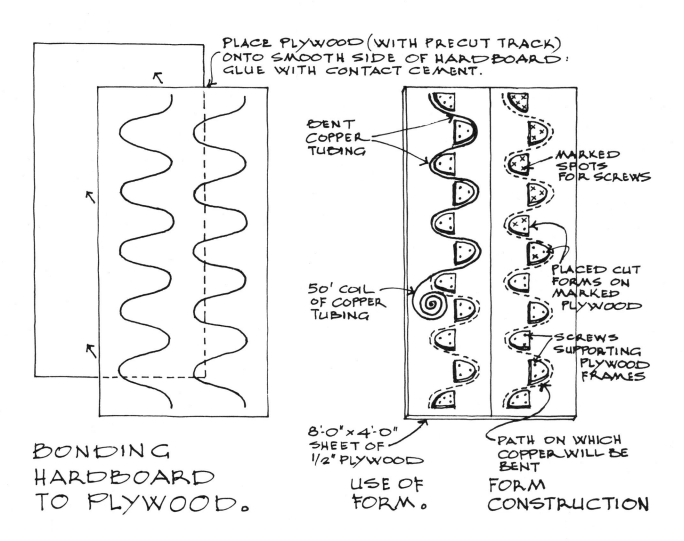

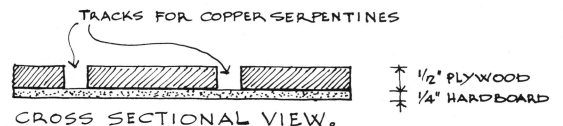

Pool Heater

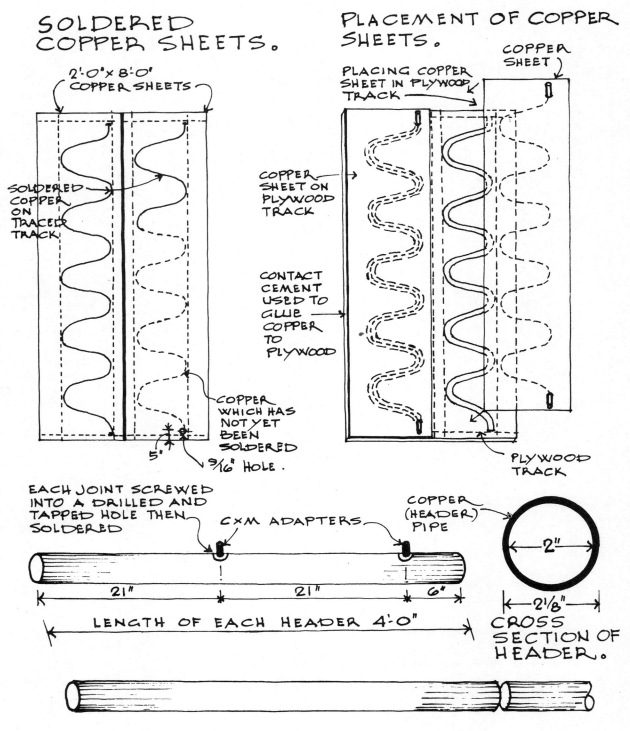

Pool Heater

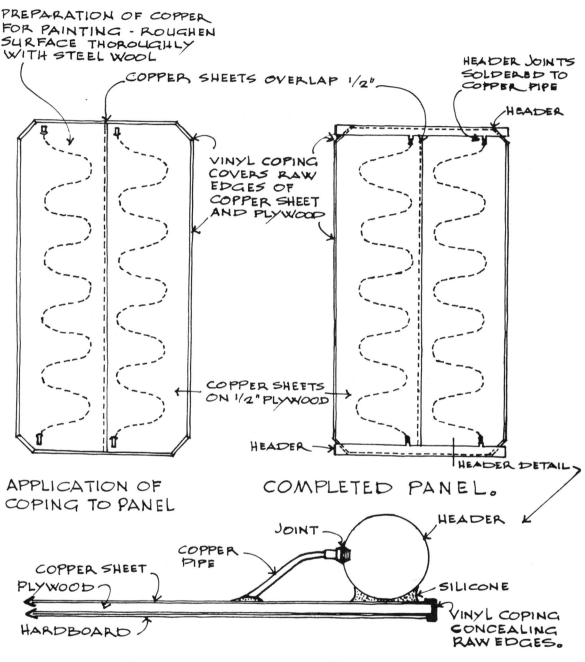

the inevitable problem of galvanic action, which can result in sacrificial dissolving of the aluminum. In addition, the problem of satisfactorily bonding the copper to the aluminum is hard to solve in a do-it-yourself system. Iron was not used because it also presents potential corrosion problems over the long term.

Each 8-by-4-foot panel consists of copper sheet with a header at either end and two serpentine riser tubes linking the headers. The tubes are soldered to the underside of each sheet. The copper is then bonded to a plywood frame and painted flat black to absorb sunlight. The individual panels are then linked together and plumbed into the pool's existing plumbing.

Design Considerations

In determining the overall design of the pool heating system a number of factors have to be taken into consideration. Using the specific system first built as an example: The volume of the pool was 27,300 U.S. gallons. The size of the pool was 18 by 36 feet, with a surface area of 648 square feet. The existing pump was a ¾-h.p. Jacuzzi. It was desired to achieve a swimming temperature of 80°F. The wind factor was not considered to be a major problem in the particular location. The angle of the roof was 20°. The optimum angle for summer use = latitude − 15° = 50° − 15° = 35° for the particular latitude in question. (N.B. Tilt can be up to 20° from optimum and 90 percent of maximum insolation is still received.)

Design calculations for pool described*

a) Volume of the pool = 27,300 U.S. gallons

b) Circulation pump circulates approximately one pool volume every 10 hours (typical). Flow rate through total solar heating system = $\frac{27,300}{600}$ = 45.5 gallons/minute (G.P.M.).

c) The number of tubes = 2 per collector. On 10 collectors there are 20 tubes.

d) Tube size = ⅜ inch nominal diameter (½ inch external diameter)

e) G.P.M. per tube = $\frac{45.5}{20}$ = 2.275

f) Flow length = 15 feet + 9 inches for each bend (9 inches added on to compensate for frictional loss due to bend; with 10 bends, therefore, 7.5 feet extra). Equivalent flow length for each riser = 22.5 feet

Solar Swimming Pool Heater

g) Pressure drop = pressure drop per foot × length
    (see accompanying graph)
    = 0.175 × 22.5
    = 3.9375
    = 3.94 lb./in.^{2} (correct to 2 decimal places)

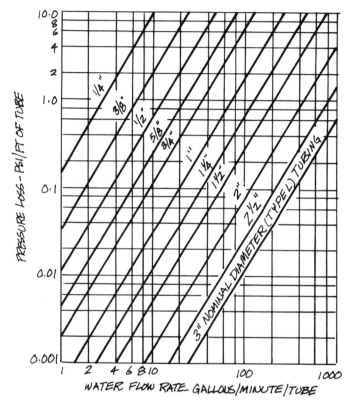

HOW PRESSURE DROP PER FOOT OF TUBE VARIES AS A FUNCTION OF WATER FLOW RATE AND TUBE DIAMETER *

*OPTIMUM SPACING OF RISERS ON COPPER SHEET.

Copper Sheet Description		Optimum Spacing of Collection Tubes	
Weight	Thickness	Unglazed	Glazed
10 oz.	0.0135 in.	7¾ in.	8½ in.
14 oz.	0.0189 in.	8¼ in.	9¼ in.
16 oz.	0.0216 in.	8½ in.	9½ in.
20 oz.	0.0270 in.	8¾ in.	10 in.

Design Guidelines

In designing a heater for your own specific pool carefully consider the following:

Pool Plumbing—If you have a small or medium-size pool, use 1½-inch tubing for the plumbing, if the pool is large, use 2-inch tubing, if it is very large, you should determine the diameter necessary.

Collector Tube Spacing—With thin sheet material, you might use a spacing of 8 inches, for thick material either 8 inches or 10 inches, depending on the width of sheet material you are able to buy. Use tubes of ⅜-inch nominal diameter, i.e. of ½-inch actual outside diameter.

Number of Tubes—Do not use less than about 10 heater tubes in parallel. If you use relatively few, use ½-inch tubing (i.e., ⅝-inch actual outside diameter). If you use about 15 tubes or more you might use ⅜-inch tube (i.e., ½-inch actual outside diameter).

Length of Tubes on Collector—On a combination heater/roof the length of the tubes should not be more than about 30 feet. On a heater which does not have a roofing function you can build it longer if you wish, but you should allow for thermal expansion, and run through some pressure-drop calculations.*

Following the guidelines outlined should prevent any potential disasters in your own system.

* Courtesy of Copper Development Association Inc.

Construction

Sequential sketches illustrating the construction of each individual swimming pool heater panel follow. To facilitate construction, the description and sketches should be followed simultaneously. To save time, it is advised that all panels be constructed at the same time rather than individually.

Laying Out Serpentines

On an 8-by-4-foot sheet of exterior-grade plywood, very carefully draw the twin serpentine arrangements, as illustrated. The accuracy of each serpentine can be checked by using a piece of string 16 feet in length. A 50-foot length of soft copper ½-inch-O.D. (⅜-inch-I.D.) can later be conveniently cut into 16-foot lengths with some room left for error.

Cutting Out Serpentines

Once the 16-foot-long lines of the serpentines have been drawn, take a compass set open at 2 inches and mark out two lines, one on either side of each serpentine line, so that two tracks, 4 inches wide, are produced. Using a saber saw, cut out the two 4-inch-wide serpentine tracks by following the lines drawn. These tracks will later accommodate the copper tubing. (The first ½-inch plywood sheet marked and cut out may now be used as a guide for marking all future serpentine tracks.)

Laying Out Serpentines on Copper Sheet

Place two 2-by-8-foot sheets of 0.0135-inch copper under the serpentine tracks. The copper sheets should overlap ½ inch and thus be ¼ inch away from each side edge of the plywood. Using a felt-tipped pen, mark out the serpentine tracks on the copper ready for later soldering.

Bonding Hardboard to Plywood

Roll contact cement onto the plywood and also onto the smooth side of an 8-by-4-foot sheet of ¼-inch hardboard. Make sure that plenty of cement is used. When the cement has dried somewhat, test to see if a piece of paper will stick to it; if it won't it is ready for use. Firmly press the hardboard to the plywood to make a good bond. Walk over the panel to ensure complete adhesion.

Forming the Copper Tubes

Next, a form is made in order to bend the copper pipe required for each panel. The form can conveniently be made on a 2-by-8-foot sheet of ½-inch plywood. Mark out a serpentine track identical to that originally drawn. Then mark and cut pieces of ¾-inch plywood to produce the individual semicircular pieces of the overall form. Bear in mind that these individual pieces of plywood must be cut ¼ inch away from the serpentine (N.B. the serpentine line represents the midpoint of the ½-inch tube). The individual forms should be firmly glued and screwed in place.

Twenty 16-foot lengths of ½-inch-O.D. copper tubing are then bent into shape, taking care not to kink the tubing. The tubing is then cut to length with a tubing cutter. The end of each tube is carefully cleaned to remove burrs and the ridge left by the tubing cutter. Two or three people can bend the tubing together much faster than one person. Surprisingly, it only takes about two minutes to bend each tube!

Headers

Using 2-inch low-pressure copper drainage pipe, cut twenty 4-foot pieces of copper that will be the headers on each collector. Try to borrow a large tubing cutter from a cooperative plumbing shop; a tubing cutter is easier to use than a hacksaw and produces a more professional finish. Remove burrs and ridges at either end of the tubes. With a centerpunch, mark spots 6 inches and 27 inches from one end of each header. Then carefully drill holes 9/16 inch in diameter and thread each hole with a ⅜-inch tap to accommodate a CxM adapter, the C opening being ½ inch I.D. After thorough cleaning, the adapters are fluxed with a good-quality paste flux, screwed into position, and soldered with a 50:50 solder. The procedure is repeated with all the other headers.

Soldering the Copper Tubes to the Copper Sheets

Next the copper tubes are soldered to the copper sheets. The sheets are first drilled to produce two holes, 9/16-inch in diameter, 5 inches from the edge of the copper (as illustrated).

The copper is then cleaned with steel wool and fluxed along the track previously marked out with a felt-tipped pen. The copper tube is then placed on the fluxed track after it also has been cleaned and fluxed. The tube is held in place by means of two iron tubes 2 feet long and 1 inch square in cross-section. The iron tubes (or equivalent) are clamped in place across the copper tube and the section in between the pipe braces is soldered with 50:50 solder. The soldering is then continued section by section. Remember to feed the ends of the tube through the holes in the copper sheet prior to soldering close to either end of the sheet.

Installing the Copper Sheets

Once the copper tubes are soldered in place the copper sheets and tubes are ready to be placed in the plywood tracks. Contact cement is applied to a 4-inch margin around the perimeter of the underside of each copper sheet and the corresponding areas on the plywood pieces with the tracks cut out. Firmly bond the copper and plywood together once the contact cement is dry. Remember to offset the 2-by-8-foot sheets so that they overlap by ½ inch at the center. The middle area of each 2-by-8-foot copper sheet is not glued to allow for a certain degree of expansion and contraction of each copper sheet and tube.

The four corners of each panel are cut off with a saber saw, using a fairly fine-toothed blade. Then ¾-inch heavy-duty vinyl window coping (U-configuration) is applied to all the edges of each panel, using clear silicone, and to ensure a tight seal a few rough-galvanized large-headed nails are also used to secure the coping. Silicone all areas of the panel that might conceivably leak at some future date; better safe than sorry!

Solar Swimming Pool Heater

Installing the Headers and Risers

Place the headers in position, clamp to the panel and solder the risers to the connectors in the headers. Caulk where the header touches the collector along both sides of the header (see illustration). Where the copper risers come through the copper panel to meet the headers, apply further silicone. Leave the silicone to dry for at least twenty-four hours before removing the clamps holding the headers in place.

Painting the Collector Panels

Roughen the surface of the copper with emery cloth, making sure that all the surface is properly cleaned and roughened. Apply a high-quality flat black paint to the panel. (A heat-resistant paint is probably unnecessary since very high temperatures are not experienced with this type of collector. I have found Rustoleum flat black paint to be quite satisfactory.) The back of each panel is thoroughly painted with cold tar to ensure that it will be impervious to water. Several coats may be necessary.

Mounting the Panels

The individual panels are mounted so that all the top and bottom headers of each panel are in line with one another and are absolutely horizontal with reference to the ground. A chalk line may be conveniently snapped on the roof or mounting rack to achieve this. The collectors should be as high on the particular roof as possible to prevent snow build-up above them.

The specific mounting instructions are as follows: six $3/8$-inch-diameter holes are drilled through each panel; two at the top, middle, and bottom of each panel, away from the internal piping.

The holes are elongated in a horizontal direction to allow for expansion or contraction of the mounted panel. Silicone or polysulphide sealant is injected into the holes and the lag bolts are firmly screwed into the roof. On a shaked roof, or wherever considered necessary, the panels may be raised from the roof surface by mounting them on three wooden slats running parallel to the headers. The slats are bolted to the roof before the panels are in turn attached to them.

The panels are set about 4 inches apart and are linked to the next panel by means of high-grade rubber piping that fits snugly over the ends of the headers. At least $1\frac{1}{2}$ inches of the rubber should cover the protruding ends of each header. The rubber hose is secured by the use of two stainless steel clamps at each end.

The rest of the plumbing is completed as per enclosed diagram, using schedule 40 or 160 plastic piping.

Low-Cost Air Heater

This air heater utilizes scrap printing plates for the fabrication of low-cost collector plates. The collector plates consist of corrugated aluminum over polished flat aluminum. The upper corrugated surface is painted matte black and absorbs sunlight that falls on its surface, heating it. Air flowing over the plates and through multiple holes in their surface then flows under the plates and out of the collector. The flow of air above and below the plates helps to minimize loss of heat from the collector and to maximize its efficiency.

Materials Required

1 4-by-8-foot sheet of ¾-inch plywood

1 2-by-8-foot piece of 3/4-inch plywood

8 3-by-2-foot, 0.009-inch used aluminum printing plates

Suggested glazing

Outer, Kalwall premium or tempered glass, 26½-by-1 90½-inches; Inner, Tedlar 3-by-8-feet (to be trimmed)

Galvanized metal

1 4-by-8-foot sheet (only if back of collector is to be flashed)

General

1 20 foot piece 9-inch-wide metal strip

2 boxes of 3-by-3-by-32½-inch (open at one end)

1 20 foot piece of fiberglass, 16 inches wide, 4 inches thick (R-12)

1 caulking cartridge of clear silicone sealant

2 cans of flat black heat-resistant paint

½ pound of 1¼-inch galvanized nails

3 dozen ½-inch aluminum screws

½ pound of ⅜-inch sheet metal screws

waterproof glue

paint

Materials The major materials required for this project are ¾-inch plywood, 4-inch-thick fiberglass batts, eight 3-by-2-foot used printing plates (0.009 inch thick), galvanized flashing, and glazing materials as described.

Collector Box From ¾-inch plywood cut out the base of the collector box, a rectangle 96 by 32½ inches. Cut two pieces of ¾-inch plywood 7¼ by 96 inches. Nail and glue these two pieces of plywood onto the top of the base, along its edges. Cut two pieces of ¾-inch plywood 5¾ by 31 inches; glue and nail these air intake and exit dividers in place 6 inches from either end of the collector box.

Intake and Exit Boxes Either fabricate, or have made, two oblong galvanized boxes, 3 by 3 by 32½ inches, each having one end open. Mark and cut out holes to accommodate these boxes; use the actual boxes as guides to mark out holes in the plywood sidepieces.

Use sheet-metal screws to attach both intake box and exit box in their respective places. Make sure the open end of each box protrudes 3 inches beyond the edges of the collector.

Collector Plate Supports Rip out, from ¾-inch plywood, side collector plate supporters 5¾ by 82½ inches and glue and nail them to the base and intake and exit dividers so that the inward-facing surface of the plywood is exactly 3½ inches from the outer edge of the collector box.

Cut five collector plate supporting struts out of ¾-inch plywood, each one 2 by 25½ inches. Attach one of these to the air intake divider and one to the exit divider so that the upper edge of the struts is 4 inches from the inner surface of the collector's base. Glue and nail the other plate supporters at equal intervals inside the collector box.

Insulation Pack the whole of the interior of the collector's structure with 4-inch-thick batts of fiberglass, making sure that all areas to be insulated are thoroughly stuffed. (Don't pack the fiberglass too tightly, however, or you will reduce its insulating properties.)

Low-cost Air Heater

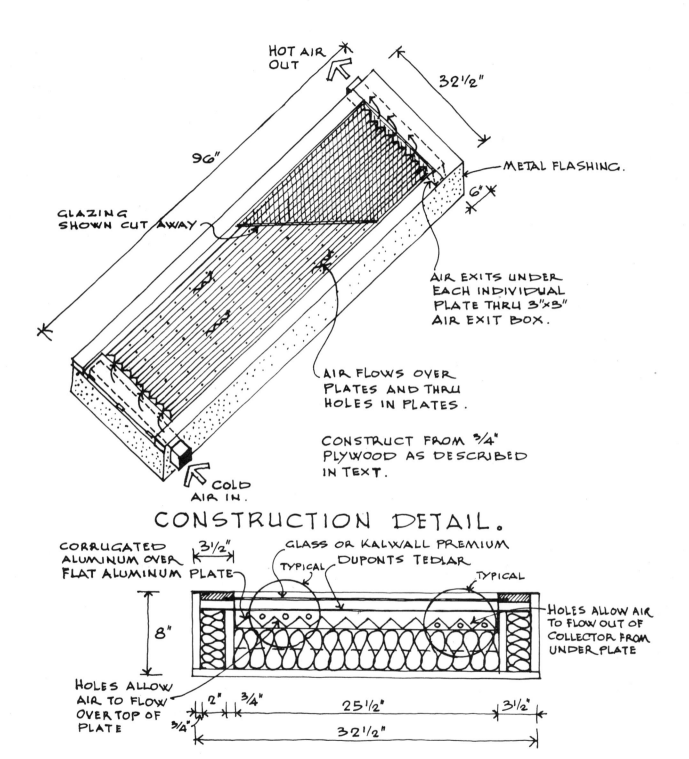

Collector Plates   Take eight printing plates 3 by 2 feet in size, and clean them with white spirits. Bend V-shaped corrugations parallel to the 2-foot sides of four of the cleaned plates; each side of each V should be approximately 1½ inches.

Halfway up each V, punch a hole with a 1-inch finishing nail through the aluminum. Repeat on either side of the V's at 4-inch intervals along the entire length of the plates. These holes will later allow airflow from one side of the plates to the other.

Assembling the Collector Boxes   Drill ½-inch holes from the interior of the collector box into the air intake box and into the exit box, bearing in mind that air goes into the collector above the plates and exits below the plates. Use the corrugated shape as a guide in determining the positions of the holes. Glue and screw the two ends of the collector box (31 by 6½ inches, made of ¾ inch plywood) into place.

Reflector Plates   Fix four flat plates of shiny polished aluminum with aluminum screws and silicone, with the 3-foot length across the box. Trim after bending and securing these reflective plates to the inner sides of the collector plate supporters.

Finishing and Installing Absorber Plates   The corrugated absorber plates should be sprayed on both sides with "hibachi flat black" or an equivalent heat-resistant paint. When the paint has dried use more screws and silicone to fix the painted corrugated sheets above the flat plates, making sure that 1½-inch strips are left on either side. These are then bent up the sides of the collector side supporters and are fixed in position. Overlap the plates as they are placed in the collector.

Glazing   Depending on the glazing to be used, either follow the drawing details (for an inner cover of Tedlar and outer glazing of Kalwall) or adapt them according to whatever other materials are available.

Finishing the Collector Box   The whole collector box can be painted or preferably clad with galvanized iron sheeting. Use clear silicone to thoroughly caulk all areas where leaks could occur.

Food-Producing Devices

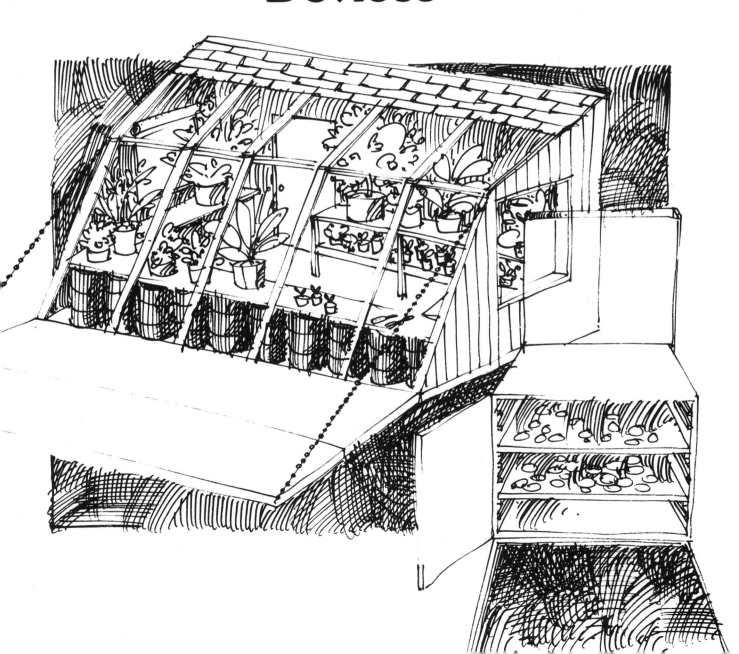

Solar Cold Frame

In the spring this solar cold frame can provide developing plants with at least two months of extra protection from possible frost damage.

Light passing through the clear (translucent) fiberglass cover of the cold frame heats water-filled sealed cans which warm up during the day. At night the closed cover, the insulation in the walls of the cold frame and the roll-over insulated blanket help retain heat and prevent frost damage. The translucent cover can be propped up to vent away excess heat during particularly warm weather. The roll-over blanket probably is not necessary except when frost is actually anticipated.

Again, ¾-inch plywood (preferably pressure-treated) is used for the majority of the construction of this project.

Materials Required

2 4-by-8-foot sheets of pressure-treated ¾-inch plywood

1 10-foot-long piece of U.V.-inhibited clear corrugated fiberglass

220 used quart-size oil cans (half to be used to obtain tops)

8 2-by-3-foot used aluminum printing plates

90 feet of 1-by-2-inch redwood or cedar

1 insulating blanket made from fabric and 4-inch fiberglass (finished blanket 46 by 44 inches)

1 4-by-8-foot sheet of 2-inch Styrofoam

1 4-by-4-foot sheet of 2-inch Styrofoam

2 gallons of antifreeze

1 caulking tube of clear silicone sealant

1 can of flat black paint

2 3-inch butt hinges

2 dozen 3½-inch screws

1 quart of cold tar

1 pound of 1¼-inch galvanized nails

18 feet of self-adhesive weatherstripping

Styrofoam glue

waterproof glue

½-inch staples

Box Construction   Cut out a base 49½ by 46¾ inches, then cut out the two sidepieces (as illustrated); base length 49½ inches, sides 18 inches and 30 inches. The top edge should be approximately 51¼ inches. Glue and nail sides onto the base.

Using ¾-inch plywood, cut out the front end, 18 by 45¼ inches, and back end, 30 by 45¼ inches. Glue and nail the ends in place.

Cover Supporting Frame   On the inside of the top of the box, fabricate a frame that will support the cover. This should be made from 1-by-2-inch cedar or other weather-resistant wood. Two pieces of 1-by-2-by-49¾-inch cedar should be angled at either end and then glued and nailed in place. Then cut two pieces 1 by 2 by 43¼ inches and glue and nail in place on the top of either end of the cold frame. Apply self-adhesive weatherstripping to the opening of the completed box to prevent possible air infiltration.

Insulation   The whole of the interior of the cold frame should be insulated with 2-inch-thick Styrofoam. The Styrofoam is marked and cut with a long-bladed utility knife. Use readily available Styrofoam glue to secure the Styrofoam in place. If the Styrofoam is cut slightly oversize, friction will help hold it in place while the glue sets.

Blanket Supports   Two blanket side supports should be cut from 1-by-2-inch cedar, 44 inches long. These should be secured in place with thin 3½-inch screws driven into the outside of the box. Five movable slats made of 1-by-2-inch cedar are cut 41¼ inches long.

Aluminum Liner   The whole of the interior is then lined with aluminum printing

Solar Cold Frame

SECTION.

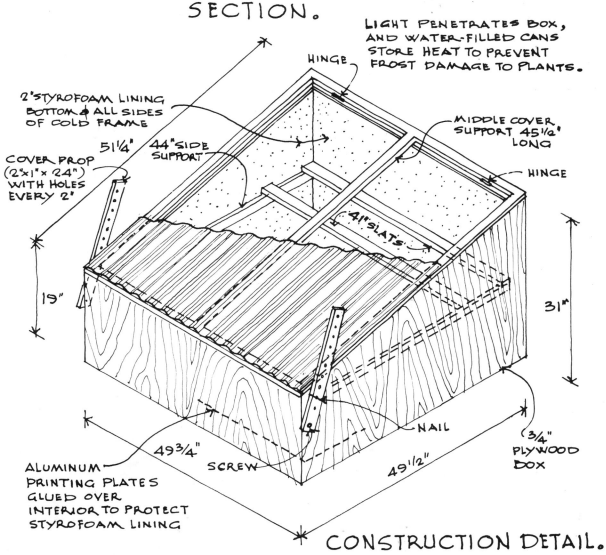

CONSTRUCTION DETAIL.

plates to protect the Styrofoam lining. Fix the plates in place with long staples and Styrofoam glue.

Cover Frame and Glazing

The top cover frame is made from 1-by-2-inch cedar fabricated as shown in the illustration. A middle cover support 47¼ inches long is cut and then glued and nailed into place in order to brace and strengthen the frame.

A 10-foot piece of standard clear greenhouse-quality corrugated fiberglass is necessary to complete the fabrication of the cover. Cut the sheet of fiberglass in two and fix into place using large-headed ring nails. Ready-made corrugated wood supports under either end of the cover, together with an ample quantity of clear silicone, completely weatherproofs the cover.

Two strong hinges are fixed to the rear of the cover frame and two hooks and eyes to the front to hold the cover and frame tight against the weatherstripping. Two cover props are made from 2-by-1-by-24-inch cedar which have ⅛-inch holes drilled in them at 2-inch intervals. They are then screwed in position onto the exterior of the cold frame. Nails hammered into the cold frame box as pegs make the props fully adjustable.

Blanket

A blanket is made from sheeting sewn into a pillow shape approximately 46 by 44 inches. This is stuffed with 4-inch-thick fiberglass batts and is then sewn up. The completed blanket is then rolled into its stored position at the back of the cold frame.

Storage Cans

About 110 one-quart cans (automobile oil cans) are necessary to fill the bottom of the cold frame. Each can has one end removed and is 90 percent filled (a gap is left to allow for expansion and contraction) with water and antifreeze and is then resealed by siliconing a top removed from another can. After thoroughly cleaning the tops of the cans, spray them with a heat-resistant flat black paint.

Waterproofing

The exterior of the cold frame is waterproofed by painting it with cold tar or an equivalent waterproofing material.

Installation

The whole cold frame is embedded in the ground to a depth of about 10 inches and your seeds are ready to go!

Model Solar Greenhouse

The solar greenhouse illustrated is a conceptual design that has not as yet been completed in full-scale form. However, basic design features to be incorporated into the full-scale version of this greenhouse are outlined below.

Conventional greenhouses are very adequately illuminated during daylight hours, but unfortunately their structures do not normally encourage good heat retention. During very cold weather greenhouses usually suffer massive heat losses and the maintenance of proper growing temperatures requires large quantities of expensive fuel.

A solar greenhouse provides an economically justifiable alternative to the typical greenhouse. In designing it, two somewhat opposing factors have to be considered: allowing the entrance of sufficient light for healthy growth while minimizing heat losses. Through its construction a solar greenhouse is in effect a combination solar collector and heat storage area. It not only serves as an area to grow food and provide controlled humidity for the adjacent home, but also provides supplemental heat.

Materials Required

Exterior and framing
Cedar or redwood

Thermal Storage
Water in used 45-gallon fiberglass resin drums plus 1½-inch-diameter rocks (under greenhouse)

Glazing
Outer, Kalwall premium

Inner, Tedlar or clear heavy-duty vinyl (not as durable as Tedlar)

Ventilating structures
Fan plus two large opening windows

Reflective panels
4 inches of Styrofoam covered with polished used aluminum printing plates

Structure   The specifics of the proposed design are as follows. Ideally, the greenhouse is built onto the south-facing wall of a house. The whole structure of the greenhouse is insulated to R-20 in the walls and R-28 in the roof. It has windows on its south-facing front, half of its roof and part of each of its west and east sides.

To avoid spindly growth because of the reduced areas of glazing, the whole of the interior of the greenhouse, apart from heat-storing structures, is painted white to reflect as much light as possible. Lights are used to extend daylight hours, and during particularly cold weather all glazing is covered and growth is maintained solely by using electric light rather than natural illumination.

A number of features help to retain heat and serve to control the intensity of light entering the greenhouse. The side windows have roll-down insulating blinds fabricated from canvas which sandwiches fiberglass batting. The inward-facing side of the canvas is finished with a thin film of reflective Mylar. Velcro fasteners around the edges of the blinds seal them to the window frame.

Reflecting Panel   A reflecting panel, faced with polished aluminum printing plates, boosts the light entering the south window during the winter, early spring, and late fall. If the intensity of light becomes too great, then the reflecting panel may be folded in two to eliminate its reflective function. The whole panel could be opened and closed by utilizing an electric motor linked to nylon ropes attached to the panel. At night the insulated reflective panel is closed and helps to reflect heat back into the interior of the greenhouse. In summer possible overheating of the greenhouse is reduced by using roll-down bamboo or vinyl shades over the south-facing windows; the side windows could also be opened.

The thermal mass of the greenhouse should be as high as possible in order to absorb and retain heat for as long as possible.

Storage Drums   Used fiberglass resin or oil drums, filled with water to 90 percent of capacity (an air space is left to allow for expansion of the

Solar Greenhouse

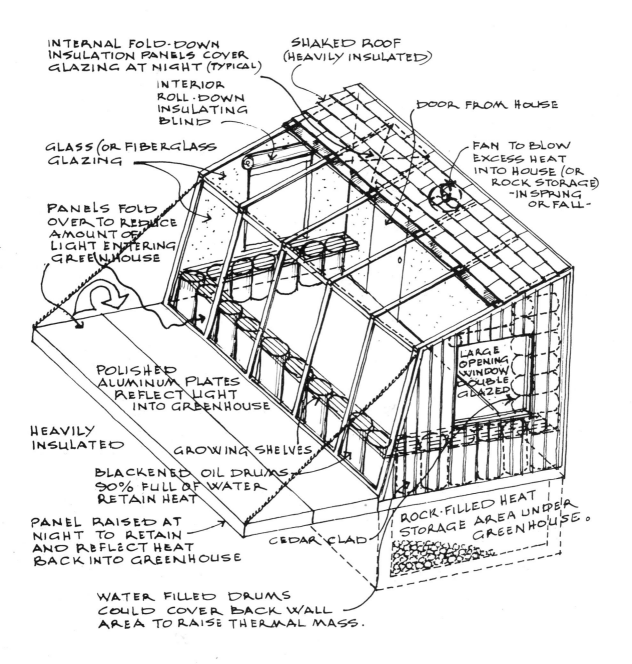

water), are placed upright around the perimeter of the greenhouse and are stacked horizontally at the back of the greenhouse. These drums, which are painted black on the side facing the sun and white on the opposite side, absorb and store large quantities of heat. In addition, the upright drums support growing shelves and trays.

Fan A fan is situated high in the greenhouse. Whenever there is hot air in excess of that required in the greenhouse it is either blown into rock storage below the greenhouse, into the adjacent home, or out of the structure altogether.

A solar greenhouse based on the above design parameters is presently under construction.

Food Dryer

There is nothing more satisfying than being able to chew on some tasty pieces of dried fruit in the middle of the cold winter months. This is a project that when completed will operate largely from solar power but can easily be adapted to use electrical drying should the need arise.

Fruit can be kept in your refrigerator for a few days until the weather is settled and then you can go into production; keep your fingers crossed!

The flat-plate collector heats up as direct or diffuse sunlight falls on it. Air inside the collector rises as it is heated and travels into the bottom front of the food-drying compartment. The hot moist air exits through a chimney which is blackened on the exterior to heat the air inside it and thus accelerate its elimination from the food dryer.

The food dryer is largely made from ¾-inch plywood.

Materials Required

- 2 4-by-8-foot sheets of ¾-inch plywood
- ½ 4-by-8-foot sheet of ⅛-inch hardboard
- 20 feet of 2-foot-wide nylon mesh screen
- 80 feet of ¾-by-¾-inch pine
- 2 2-by-3-foot used aluminum printing plates
- 1 24-by-18-by-3-inch light-gauge galvanized chimney
- 1 4-by-23¾-inch light-gauge galvanized draft controller
- 1 24¾-inch piano hinge
- 4 3-inch steel brackets
- 1 2-by-4-foot sheet of heavy-duty clear vinyl
- 1 thermometer
- 1 can flat black heat-resistant paint
- 1 quart high-gloss clear polyurethane paint
- 1 hook and eye
- 8 feet of self-adhesive weather stripping
- ½ pound of 1½-inch galvanized finishing nails
- ½ pound of 1¼-inch galvanized finishing nails
- paint for exterior
- waterproof glue

Food Dryer

Drying Compartment The box which is to be the food-drying compartment is first constructed. Cut up plywood to the following dimensions: the door 24 by 24¾ inches; the back 24 by 24 inches; the two sides 24 by 22½ inches; the top 24 by 20 inches, and the bottom 22½ by 19¼ inches.

The two sidepieces should be glued and nailed to the back; nail through the back into the 24-inch lengths of the two sides. The top is then glued and nailed onto the top of the partially completed box, making sure to leave an opening for the chimney duct at the rear of the dryer. The bottom is then glued and nailed inside the hole left at the bottom of the dryer. Ensure that a 3¾-inch duct opening is left where the collector will later be attached.

Supporting Slats Do not put the door on the dryer compartment until all the runner slats are attached. Cut ¾-by-¾-inch pieces of pine 22½ inches long and glue and screw these slats into place, leaving a 1-inch gap at the bottom of the dryer before the first supporting slat is put in place. Thereafter leave a 1⅜-inch gap between each of the supports; there are ten supports on either side of the dryer.

Drying Trays Next comes the most the most tedious job—the construction of the drying trays.

Cut twenty pieces of pine ¾ by ¾ inch by 22⅛ inches long and twenty pieces of the same stock 20⅝ inches long.

Glue and nail the pieces together with the 20⅝-inch pieces fitting in between the ends of the longer 22⅛-inch pieces. Make sure that the completed frames are square. Stretch nylon mesh screen over the frames and staple it in place. Rip 20 pieces of ⅛-inch hardboard 22⅛ inches long and ¾ inch wide and 20 pieces 20⅝ inches long and ¾ inch wide. Glue and nail the 22⅛-inch-long pieces over the wooden frames, permanently sandwiching the mesh in place and also covering the joints in the underlying pine, thus giving the whole frame a lot more strength. The frames of the drying trays and the interior of the dryer may be sealed with urethane to produce a hard, washable, permanent finish.

Installing the Door Using a piano hinge, attach the front door to the dryer. Weatherstrip the opening with self-adhesive foam weatherstripping. Screw in a hook-and-eye fastener to hold the door firmly in place.

Leg Supports The legs are made from four pieces of ¾-inch plywood 2 inches wide; two of the legs should be 54¾ inches long and the other two 72¾ inches. Glue and nail these legs to the sides of the dryer as illustrated. Cross-brace the legs with further pieces of plywood, 2 inches wide; two being 24 inches long and the other two 25½

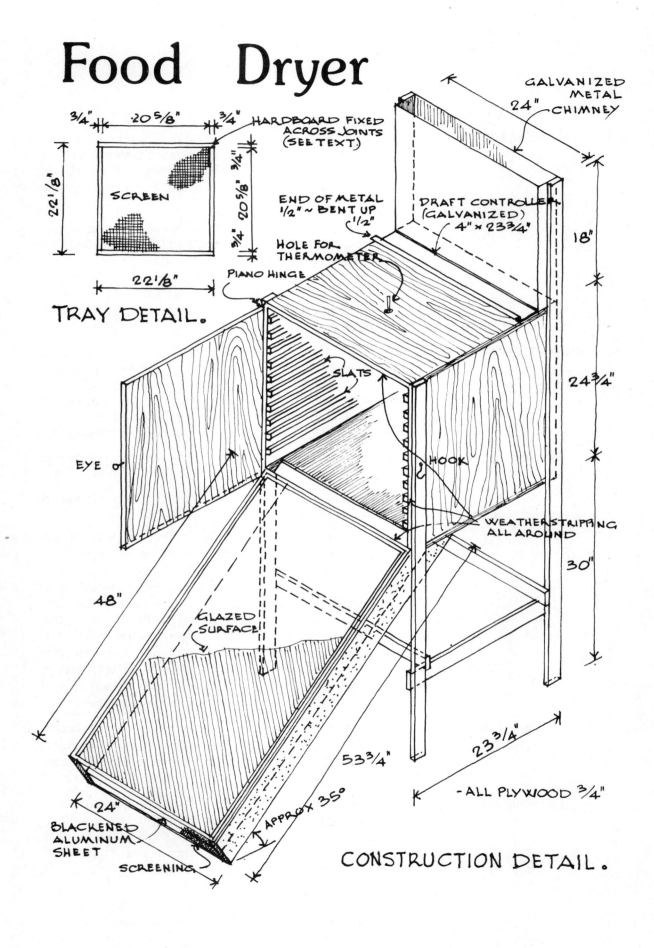

inches long (see illustration). Brackets may be attached to each leg and the base of the dryer if desired.

Chimney The chimney is fabricated from light-gauge galvanized sheet metal; its overall dimensions should be 18 by 3 by 24 inches. However, the back edge of the lower end of the chimney should extend an additional inch in order that the chimney may be firmly screwed to the back of the dryer. The extended legs are screwed to the chimney. A gap of 1/16 inch should be left between the chimney and the main body of the dryer to accommodate the draft controller. The chimney is then sprayed with heat-resistant flat black paint.

Collector The collector basically consists of a box which is single-glazed on the side facing the sun and contains a suspended sheet of blackened aluminum that heats up as the sun strikes it. The collector should be 4 feet in length in order to collect sufficient heat to dry the food. A 48-by-24-inch collector may be constructed as follows.

Using ¾-inch plywood, cut a base 53¾ by 22½ inches. Then cut two sidepieces, 4 inches wide and 53¾ inches long on the bottom edge, and 48 inches long on the upper edge. Glue and nail these pieces to the long sides of the base. Cut two collector plate supporting strips of pine ¾ by ¾ inch and glue and nail lengthwise so that the upper surfaces of the strips are 2 inches from the base of the collector. Obtain two 2-by-3-foot printing plates, rub them with steel wool and clean them with white spirits. Then, between the supporting strips already in place, attach four more strips, each 2 by ¾ by 1½ inches. Place one of these supporting strips at either end of the collector, and space the other two according to the exact dimensions of the printing plates you obtain. Clean the aluminum plates well and spray both sides with flat black heat-resistant paint. Then silicone and staple the plates into position after they have been trimmed to size.

Glazing Glaze the collector with glass or plastic and hold the glazing in place with silicone and corner-profile 1-by-1-inch heavy-duty vinyl molding. Put a piece of screening across the bottom end of the collector to prevent insects from getting into your dryer. Screw the top end of the collector in position connecting it to the main body of the dryer.

You are now ready to dry up a stock of goodies for next winter's munching.

Experimental Solar Devices

Experimental Concentrating Water Heater

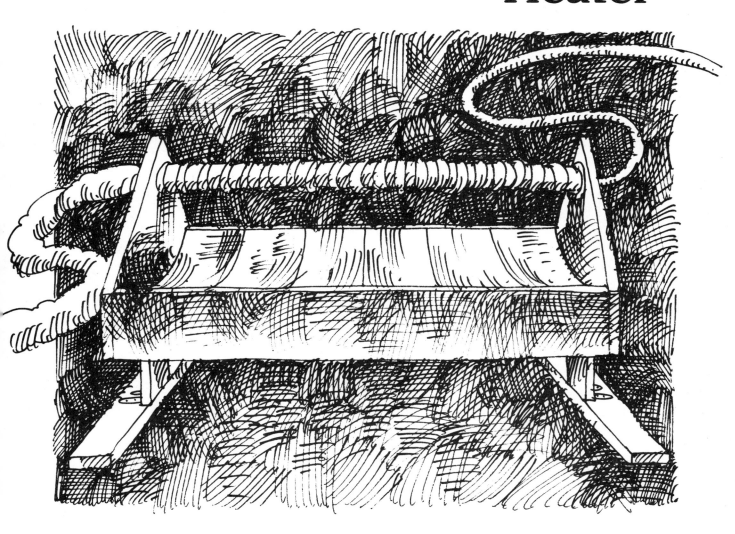

This experimental concentrating water heater was built in the course of a series of experiments to see how efficient such designs actually are. An automatic tracking system would be necessary to make this heater really practical for heating domestic hot water.

Flat-plate collectors are generally more appropriate for heating water than concentrating collectors. However, concentrating collectors do have the advantage of not requiring as much expensive glazing material as flat-plate collectors.

Light striking the highly polished aluminum surface of the collector reflects light through a rounded double-glazed fiberglass window. The light is then absorbed by a copper tube soldered to a strip of copper; both are blackened. Water flowing through the pipe absorbs heat, while the glazing and the insulation surrounding the collector tube reduce heat losses.

Materials Required

3 4-by-8-foot sheets of ¾-inch plywood

1 4-by-8-foot sheet of ⅛-inch hardboard

1 8-foot-long piece of 6-inch-diameter stove pipe (or, alternatively, thin-walled asbestos pipe)

1 8-foot strip of 6-inch wide 0.135-inch copper

1 12-foot piece of ¾-inch hard copper tubing

8 feet of 16-foot wide, 4-inch thick (R–12) fiberglass insulation

6 strips of Kalwall premium fiberglass 30-by-5½-inches

4 6-inch steel brackets

2 2-inch bolts, ½-inch diameter

1 pound of 1¼-inch galvanized finishing nails

½ pound of 1-inch plasterboard nails

1 caulking tube of clear silicone sealant

1 can of flat black heat-resistant paint

paint

glue

silver solder

flux

dilute hydrochloric acid

Experimental Concentrating Water Heater

First cut six pieces of ¾-inch plywood 10 inches wide and 48 inches long. Mark parabolas on these pieces, 48 inches wide and having a focal point of 18 inches (see instructions for the hot dog cooker). Cut out the parabolas using a jig- or bandsaw.

Cut six 2-inch-wide strips of ¾-inch plywood, 48 inches long; glue and nail these to the bottom of the six parabolas. These strips should be attached on center to four of the parabolas, the other two parabolas (which will be placed at either end of the collector) should be attached off-center, with one edge of the strip level with the side of the parabola in question.

Cut three strips of ¾-inch plywood, 96 inches long and 6 inches wide. Glue and nail the parabolic ribs of the collector to these three plywood pieces, one at either end of each rib, the other one at the center. Leave a gap of 16 8/10 inches between the base of each rib.

Cut two pieces of ⅛-inch hardboard, 24 inches wide and 96 inches long. Glue and nail these pieces to the parabolas in place, as shown in the illustration. Make sure the smooth side of the hardboard faces upwards.

To the collector attach end pieces of 3/4-inch plywood, the base of which should be 48 inches, sides 10 inches high, and above a triangular shape with its top 30 inches above the base (see illustration).

Drill ¾-inch holes in the end pieces of the collector, 20¾ inches above the center of each piece. Extend each hole 3 inches in two directions, by cutting horizontally with a jigsaw, for later admission of the collector's copper tube and copper strip. Also drill a ½-inch hole, 8¾ inches below each of the holes already drilled, to accommodate the bolts that will connect the collector to each stand.

Stands

Stands are made from ¾-inch plywood. Cut two pieces ¾ by 6 by 48 inches and two others ¾ by 6 by 44 inches. Using brackets, assemble the stands as illustrated. Join the stands to the collector assembly by placing ½-by-2-inch bolts through the holes already drilled.

Reflective Surface

The collector surface is covered with aluminum printing plates that are first cleaned with white spirits followed by dilute hydrochloric acid, using great care. Attach the plates with contact cement. Self-adhesive aluminized Mylar or strip mirrors may be used instead of aluminum plates if you want to produce a surface with excellent reflective properties.

Pipe Collector

The pipe collector assembly is fabricated next. A 6-inch

Experimental Concentrating Water Heater

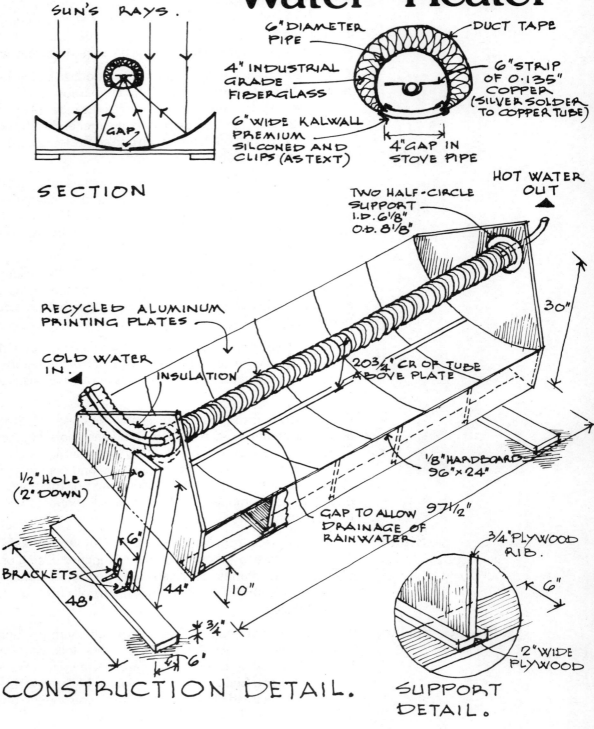

stovepipe, or an asbestos tube, is cut so that three openings are produced in line with one another, each one being 30 inches long and 4 inches wide. This leaves a 1½-inch piece of metal or asbestos before and after each opening to help strengthen the pipe collector. (Note: If asbestos is used, wear a respirator over nose and mouth to avoid inhaling the dust while cutting. Do not allow asbestos dust to contaminate your clothing, and clean work area thoroughly at once.)

Glazing Six strips of Kalwall premium 30 inches long and 5½ inches wide are bent and are then clipped in place after a liberal application of silicone under and above the glazing (see illustration). It is essential that the glazing does not touch the metal tube.

Insulation of Collector Industrial fiberglass, 4 inches thick, is wrapped around the collector pipe and duct tape is carefully wrapped right around the fiberglass and collector glazing. A ½-inch strip of galvanized metal is screwed along the edges of the glazed area to secure the tape. The duct tape is then removed from the glazing area by carefully using a razor blade. Half-circle supports of ¾-inch plywood are used to secure the collector tube in place. These supports should be 2 inches wide and have an inner diameter of 6⅛ inches and external diameter of 8⅛ inches.

Copper Tube and Strip A 6-inch-wide strip of 0.135-inch copper 96 inches long is crimped over a 12-foot straight length of ¾-inch copper pipe. The pipe is then silver-soldered to the middle of the strip.

Spray the side of the copper tube and strip that will face the glazing with flat black heat-resistant paint. Slide the copper tube and strip through the openings left in the two ends of the collector; then bend the protruding copper pipe as required.

Your concentrating water heater is ready for experimental or practical use.

Fresnel Lens Concentrator

The Fresnel lens concentrator illustrated is capable of producing temperatures in excess of 2000°F., sufficiently high to melt many common metals. It is a project whose value is largely that of providing a demonstration and producing an experimental unit rather than an object of any very practical value. The lens may be used to concentrate light onto photovoltaic cells, providing the cells are placed closer to the lens than the focal point. One has to be careful not to deteriorate the cells by overheating them.

Materials Required

1 Fresnel lens (see text)

¼ 4-by-8-foot sheet of ½-inch plywood

1 13¼-by-6-inch piece of ¾-inch plywood (firebrick support)

12 inches of 2-by-2-inch clear fir

2 1¼-by-¼-inch bolts with washers and butterfly nuts

2 2½-by-⅜-inch bolts

4 feet of self-adhesive foam weatherstripping

4 No. 10 1¾-inch wood screws

½ pound of ¾-inch finishing nails

waterproof glue

Lens

The Fresnel lens itself (which may be obtained from Edmund Scientific*) consists of a rigid plastic sheet which has multiple grooves very accurately cut into its surface. When parallel rays of light pass through the lens the light is refracted and meets to form a small intense focal point. Objects to be heated and melted are placed on a movable firebrick and support, which can be positioned closer to or farther from the lens to focus the light.

*Edmund Scientific Co.: 101 East Gloucester Pike, Barrington, NJ 37646; 3500 Bathurst St., Toronto, Ontario, M6A2C6

Lens Frame

The lens-holding frame can be conveniently made from ½-inch plywood. Cut eight pieces of plywood 14¼ by 1¾ inches; miter each corner at 45°. Four of the pieces should have grooves rabbeted along their inner edges, ¼ inch deep and 5/16 inch wide (see illustration). The bottom frame should be glued, nailed, and clamped together and the same process repeated with the top frame. Make sure each frame is perfectly flat and square. A thin

Fresnel Lens Concentrator

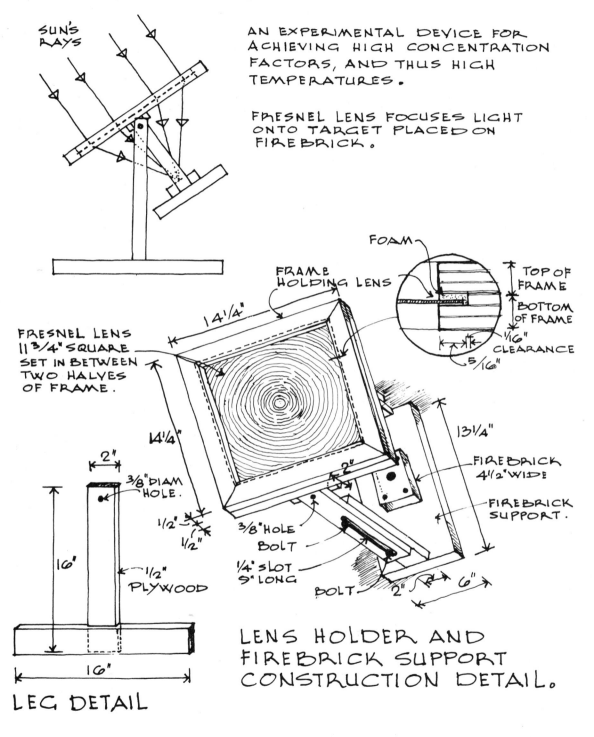

AN EXPERIMENTAL DEVICE FOR ACHIEVING HIGH CONCENTRATION FACTORS, AND THUS HIGH TEMPERATURES.

FRESNEL LENS FOCUSES LIGHT ONTO TARGET PLACED ON FIREBRICK.

LENS HOLDER AND FIREBRICK SUPPORT CONSTRUCTION DETAIL.

LEG DETAIL

strip of self-adhesive weatherstripping foam is placed opposite the ¼-by-5/16-inch groove (see illustration). The foam will hold the lens firmly in place when mounted. Eight screws may be used to fasten the two frames together. These can be withdrawn if the lens has to be removed at any time.

Lens Frame Support Arms

The slotted side arms, which support the lens frame and join it to the firebrick support arms, are fabricated next. Two 1-by-2-by-2-inch blocks are cut and each is glued and nailed to one end of two 12-by-2-inch strips of ½-inch plywood. Slots ¼ inch wide and 9 inches long are cut into the side arms. The cut for each slot should be started 2 inches from the lens frame. Follow the illustrations in conjunction with the text. A ⅜-inch hole is drilled in the slotted side supports 1 inch from the lens frame. The slotted side supports are then glued and nailed to the bottom lens frame holder.

Firebrick Support

Next the firebrick support is fabricated from ¾-inch plywood and a half firebrick (4½ by 4½ by 1¼ inches when cut). The firebrick support is cut 13¼ by 6 by ¾ inches. Then two side arms should be cut, 2 by 10 by ½ inches; set these into the middle of either end of the firebrick support. Slots should be cut into these side arms to match the slots in the frame supporting arms.

Mounting the Firebrick

Using a carbide-tipped drill, drill four ⅛-inch holes through the firebrick; ensure that each hole is at least ½ inch from the edge of the firebrick. In the middle of the firebrick a depression may be made using a ½-inch carbide-tipped bit. Fix the firebrick to the firebrick support using 1¾ inch screws. Two 1¼-by-¼-inch bolts and butterfly nuts are placed through each of the slots in the firebrick and lens support arms to make the whole structure fully adjustable.

Stand Legs

Finally, two stand legs are constructed. The base of each leg is made from wood 16 by 3½ by 1½ inches. At the midpoint of each base piece a ½-by-2-inch recess is cut to accommodate a 16-by-2-by-½-inch plywood vertical support (see illustration). Before each vertical support is fixed in position a ⅜-inch hole is drilled in the middle of the support, 1 inch from the top of the support. Glue and nail the vertical supports into place. Attach the stand legs to the rest of the concentrator already constructed with 2½-inch-long ⅜-inch bolts.

The project is then ready for very hot action!

Giant Experimental Reflector

This project is for experienced solar enthusiasts, not beginners. A giant experimental reflector was constructed so that a group of highly motivated students could build one of the largest (if not the largest) solar furnaces ever built by students of their age. Certainly this furnace has the potential to provide sufficient power to drive a Stirling engine or other engine dependent on a fairly high temperature source. The basic structure of this giant experimental reflector is essentially the same as that of the parabolic concentrating reflector described previously.

Highly polished recycled aluminum printing plates or self-adhesive aluminized Mylar provides the reflective surface for the collector. This surface can be inclined at different angles by adjusting the inclination of the swiveling collector. Rotational movement is achieved by moving the base (this movement could be facilitated by mounting the base on wheels).

The whole structure is susceptible to wind damage and therefore is intended primarily for intermittent rather than continuous use.

Materials Required

2 4-by-8-foot sheets of ¾-inch plywood

4 4-by-8-foot sheets of ½-inch plywood

1½ 4-by-8-foot sheets of ⅛-inch hardboard

48 feet of 2-by-8-inch softwood

46 feet of 1-by-1-inch softwood

50 square feet of self-adhesive aluminized Mylar mounted on hardboard or highly polished recycled aluminum printing plates

4 12-inch steel brackets

6 2-inch-long, ½-inch diameter bolts

1 11-foot heavy-duty chain

1 pound of 2½-inch rough-galvanized nails

1 pound of 1½-inch galvanized finishing nails

1 3-inch screw-threaded hook

½-inch staples

waterproof glue

paint

dilute hydrochloric acid

Furnace Base

The base of the furnace structure is made in two separate halves which are then joined when they are completed. This allows the structure to be disassembled and moved through standard-size door openings.

First start the construction of the collector itself. Place two 4-by-8-foot sheets of ¾-inch plywood side by side on a flat surface. Using a straight piece of wood 50 inches long, hammer two nails in the wood, 1 inch from either end of the wood, making sure that the nails are exactly 48 inches apart. Using this as a compass, scribe out a circle on the plywood having a radius of 48 inches.

Two pieces of ¾-inch plywood are ripped 96 inches long and 18 inches wide. Mark out on these pieces a complete parabola having a width of 96 inches and a focal point of 36 inches. (Use the method of parabola formation described earlier for construction of the hot dog cooker.)

Taking one piece of the plywood furnace base (previously marked out with the semicircle on it), attach one parabola along its edge. A 1-by-1-inch wooden strip should be nailed to the base and parabola to give additional support to the parabola. The process just completed should be repeated with the other half of the furnace base.

Reflector Shell

Using ½-inch plywood, cut thirty half-parabolas, 4 feet wide and having a 36-inch focal point. Glue and nail 1-by-1-inch strips to one side of the base of each of these half-parabolas. Then attach them radially, as illustrated, placing them 9–4/10 inches on center. Many of the half-parabolas will require cutting short on their narrow ends in order to fit on the collector base. The central

Giant Experimental Reflector

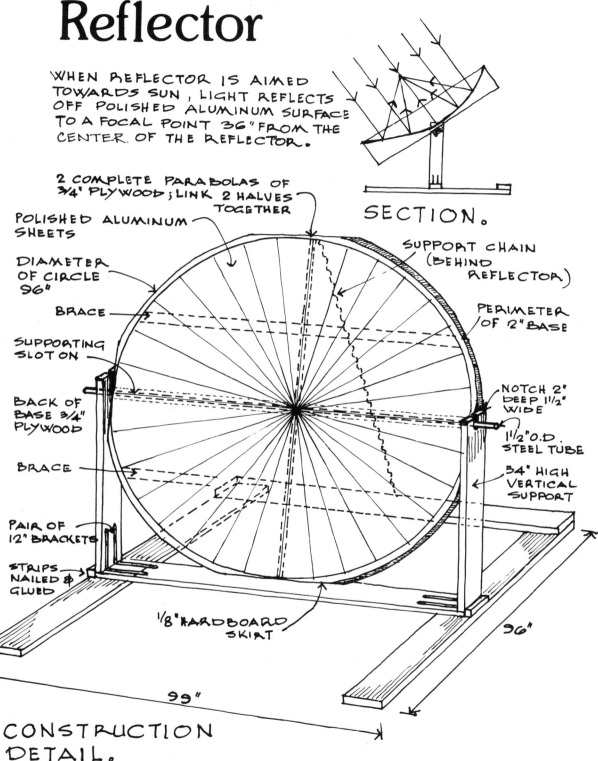

area of the collector may be filled in with wooden blocks which when attached can be belt-sanded to match the remaining parabolic shapes. Two or three 1-by-1-inch bridging struts can be glued and screwed between each two ribs to help stabilize them.

A U-shaped supporting slot, to accommodate the 1½-inch-O.D. steel tube which will support the collector and allow adjustments is fabricated from ¾-inch plywood. Cut one piece 1½ inches wide and 96 inches long, and another two pieces 3 inches by 96 inches. Glue and nail the 3-inch-wide strips to the sides of the 1½-inch strip. Reduce the 96-inch-long structure to two 48-inch-long pieces by cutting it in half. Glue and nail these 48-inch-long slot-shaped plywood structures to the middle of the back of the collector, across the joint in the furnace base, as illustrated. Make two other braces of a shape similar to the one completed and attach them to the furnace base, parallel to the slot already in place, and 21 inches away from it. Trim the braces to length.

Lining the Reflector

Cut around the circumference of the 96-inch-diameter plywood base, leaving an extra 2 inches of plywood beyond the outer circumference where possible. Thoroughly clean 3-by-2-foot aluminum printing plates with white spirits, followed by dilute hydrochloric acid. Wear rubber gloves and protective glasses and use great care when carrying out this procedure. Use the plates to cover the whole inner surface of the parabolic reflector shell. Start from the edge of the collector, gluing and stapling the plates between adjacent ribs. Then trim each plate and continue fixing them in place until the shell is completed.

Perimeter of Reflector

Hardboard, ⅛ inch thick, should be cut into 18-inch-wide strips; you will require approximately 30 linear feet. These pieces are then glued and nailed to the perimeter of the reflector assembly, i.e., to the 18-inch-wide end of each half-parabola. Leave gaps between the two halves of the furnace, between the two large parabolas, and between the first adjacent half-parabolas on either side of the two full parabolas to allow access when joining the two halves of the furnace together.

Assembling the Furnace

Place the two halves of the furnace next to each other and drill at least three ½-inch holes at either side in the central parabolas. Use ½-by-2-inch bolts to securely link the two halves together.

Support Base

The support base is made from 2-by-8-inch fir (actual dimensions 1½ by 7½ inches). Cut two sidepieces 1½ by 7½ by 96 inches. Glue and nail a rear crosspiece, 1½ by 7½ by 99 inches, to join the sidepieces together. A central crosspiece, 1½ by 7½ by 96 inches, is attached between the sidepieces of the base. This crosspiece should form a span between the midpoints of the two sidepieces.

Giant Experimental Reflector

Two 1½-by-7½-by-54-inch vertical supports are then cut; notches 2 inches deep and 1½ inches wide are cut into the top of the supports. Then glue and nail the supports in place. Supporting blocks, 1½ by 7½ by 1½ inches, are attached to strengthen the verticals (see illustration). Two 12-inch brackets are then attached to each of the vertical supports to further strengthen them.

Pivot Shaft Into the notches at the top of the vertical supports, place a steel tube 107 inches long and having an outside diameter of 1½ inches.

Adjustment Chain A stout chain 11 feet long is attached to the base of the collector and to a hook on the rear strut of the stand to make it readily adjustable.

Find many friends to help mount the collector in place; then you are ready to experiment!

Miscellaneous Solar Devices

Glass Solar Still

The all-glass still illustrated on the previous page is designed to produce distilled water even when weather conditions are not optimal.

Brackish water or tap water (*not* alcohol—it's illegal!) is placed in the central compartment. Diffuse or bright sunlight passing through the cover glass penetrates the water and glass and then hits the undercoat of tar. This absorbs sunlight, producing heat, which then helps to evaporate the water. Experimental evidence suggests that having the black layer under the glass maximizes heat output.

The evaporating water condenses on the upper glass cover and slowly but steadily drips into the two side troughs. These distilled water troughs have a layer of white silicone on their bottom surface which helps to reflect light and minimize reevaporation of the collected water. The inert nature of silicone prevents contamination of the distilled water.

The water collected is ideal for use in steam irons, for topping up lead-acid cells, for making pure chemical solutions or for drinking purposes.

The water is withdrawn when required by removing the two rubber stoppers and siphoning out the water with a rubber or copper tube. The 4-inch-diameter inspection hole is sufficiently large to permit thorough cleaning of the interior should it become necessary. Water is not lost through this inspection hole as it is well sealed by the weatherstripped glass plate covering it.

The still is almost entirely built out of glass which is joined and sealed with clear silicone. (Having built over twenty aquariums with students, I am a great believer in silicone.)

Materials Required

3/16-inch glass

- 1 2-by-3-foot base piece
- 2 24-by-20-inch top pieces
- 4 $23\frac{5}{8}$-by-3-inch trough dividers and end pieces
- 2 side pieces, 3-foot base, 1 foot high at center, 3 inches high at ends

1 7-by-6-inch inspection hole cover

1 2-by-1-inch (handle)

2 feet of 1-by-1-inch redwood or cedar

1 caulking cartridge of clear silicone sealant

2 ½-inch-diameter stoppers

1 roll of masking tape

Unless you are quite skilled at glass-cutting, I would suggest that you get that part of the job done by a glass shop. Trying to save money in this case can prove to be an expensive experiment. Used 3/16-inch glass is quite satisfactory and is probably the least expensive glass that has sufficient strength for the project.

Cutting the Glass Components

Using 3/16-inch glass, the quantity of glass required is as follows: a base piece 2 by 3 feet; two top pieces 24 by 20 inches; two endpieces and two trough dividers 23⅝ by 3 inches; and two sidepieces, the base length of which should be 3 feet, height 12 inches at the center (see illustration) and ends 3 inches high. One of the sidepieces should have a hole at least 4 inches in diameter cut in it, the center of the hole being 5 inches from the base. Two smaller holes, ½ inch in diameter, should be drilled 2 inches from the ends and 2½ inches from the base of this same sidepiece.

Assembling and Sealing with Silicone

A bead of clear silicone is then run around the top side of the perimeter of the base. The two endpieces are placed on top of the end beads of silicone. Beads of silicone are then run over the inside edges of the endpieces. (Many hands make light work here, too few make a lot of broken glass!) The sidepieces are pushed down onto the base and then moved firmly against the edges of the endpieces. Masking tape can be used to temporarily hold everything firmly together.

Next silicone the two trough dividers into place by applying the silicone to the edges of the glass that will contact the glass already in place. Each divider should be 4 inches from either end of the still. Leave the silicone to set for twenty-four hours.

White silicone is then smeared over the inner bottoms of the two distilled water collection troughs. Use clear silicone to apply a bead to all interior joints to ensure watertightness. Leave to dry for twenty-four hours. Put water in the still to check for leaks; repair if necessary.

Using a razor blade, remove all excess silicone from the surface of the glass. Fine steel wool gently applied may help to remove any remaining traces of silicone.

Silicone should then be placed all along the upper edge of all the

Glass Solar Still

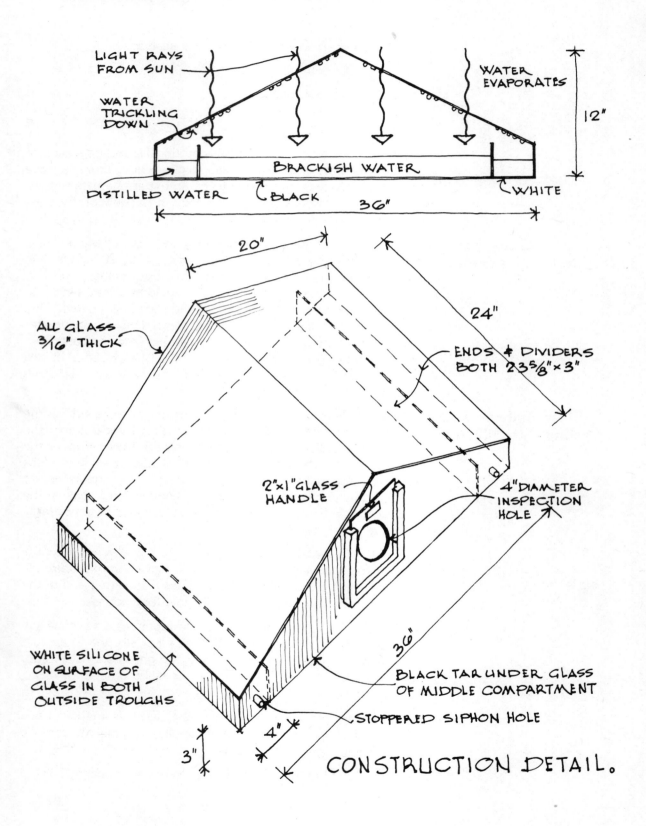

Glass Solar Still

glass in place (apart from the trough dividers). Carefully lower the two top pieces into place and tape securely into position. Remove tape when the silicone is dry and clean up excess.

Inspection Hole Cover

The final job is to fabricate the inspection hole cover. Use 1-by-1-inch cedar and cut three pieces: two pieces 6 inches long and one 7½ inches long. Cut a ⅜-by-⅜-inch slot away from one edge of each of the two pieces. Glue with silicone symmetrically on either side of the opening in the glass and butt against the 6-inch-strip. It is the ⅜-inch edge of the two vertical pieces that should be glued onto the glass.

When the silicone is dry, apply thin weatherstripping to the interior edge of each slot to hold the glass cover plate tight against the side of the still. Cut a 7-by-6-inch piece of glass to fit in the slot. Silicone a 2-by-1-by-3/16-inch glass handle close to the top edge of the inspection hole cover.

Plugs for Water Outlets

Rubber stoppers may be used as plugs in the holes from which the distilled water is drawn.

Your still is ready for action!

Solar Wood Igniter

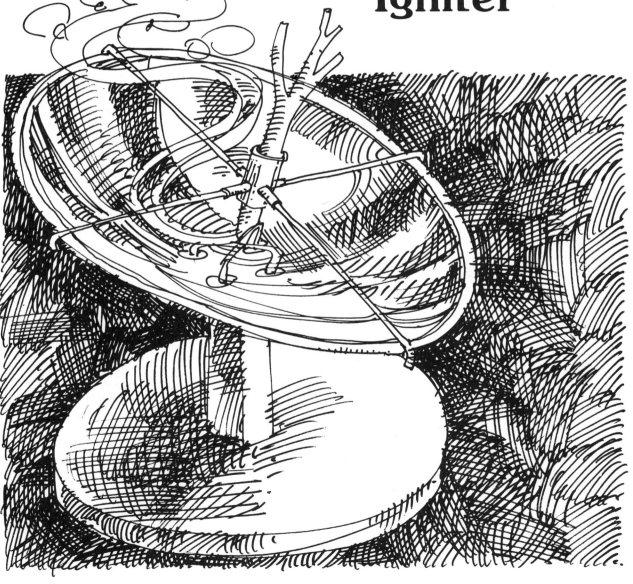

The solar wood igniter provides an interesting conversation piece, even if it is not totally practical. It is remarkably effective, capable of lighting a piece of wood within seconds.

The sealed-beam headlamp unit, minus its glass lens, is pointed towards the sun and is swiveled up and down to focus on the ends of the filament supports. The object to be ignited is lowered through a special holder to the focal point and rapid ignition follows.

Materials Required

1 used sealed-beam unit (headlamp)

1 ¾-by-9-by-9-inch piece of plywood for base

1 8-by-2-by-1½-inch piece of softwood

18 inches of ⅛-inch-diameter steel rod

1 1-inch length of 5/16-inch I.D. copper tubing

1 U-shaped steel support 2 by 2 by 2 inches (approximately; see illustration)

1 ⅜-inch-diameter 3-inch bolt with washer and butterfly nut

1 small tube glue

paint

Reflector

When you select a sealed-beam unit, make sure that the rear aluminized finish is not unduly tarnished. Wearing safety glasses, very carefully remove the lens, using a cutting blade to grind off the glass. I would emphasize that extreme care is required; this is not a procedure that should be carried out by an inexperienced person without supervision.

Once the glass has been removed as cleanly as possible, unsolder the connectors on the back of the reflector unit. This may be conveniently achieved by playing a propane torch flame on the solder. Be careful not to heat any glass, as it cracks very easily.

Reflector Support Bracket

Weld or braze together the reflector support bracket as shown in the illustration. Drill ⅜-inch holes on either side of the metal support. Drill further holes in the support to match the soldered joints on the back of the reflector unit. Turn the unit upside down and solder the steel support in place.

Base

Using a jigsaw, cut out a 9-inch-diameter base from ¾-inch plywood. Cut a 2-by-1½-inch hole in the center of the base. Cut a

Solar Wood Igniter

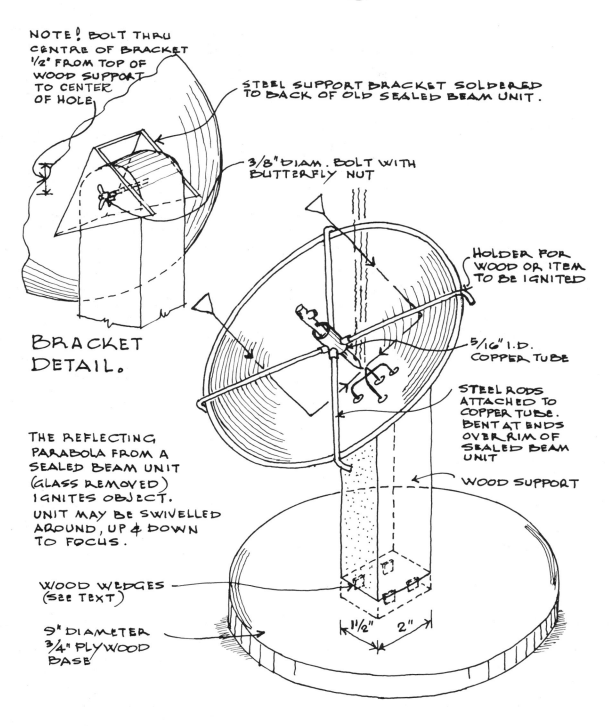

BRACKET DETAIL.

NOTE! BOLT THRU CENTRE OF BRACKET 1/2" FROM TOP OF WOOD SUPPORT TO CENTER OF HOLE

STEEL SUPPORT BRACKET SOLDERED TO BACK OF OLD SEALED BEAM UNIT.

3/8" DIAM. BOLT WITH BUTTERFLY NUT

HOLDER FOR WOOD OR ITEM TO BE IGNITED

5/16" I.D. COPPER TUBE

STEEL RODS ATTACHED TO COPPER TUBE. BENT AT ENDS OVER RIM OF SEALED BEAM UNIT

WOOD SUPPORT

THE REFLECTING PARABOLA FROM A SEALED BEAM UNIT (GLASS REMOVED) IGNITES OBJECT. UNIT MAY BE SWIVELLED AROUND, UP & DOWN TO FOCUS.

WOOD WEDGES (SEE TEXT)

9" DIAMETER 3/4" PLYWOOD BASE

CONSTRUCTION DETAIL.

piece of wood 8 by 2 by 1½ inches, rounding off the top end. Drill a ⅜-inch hole close to the top end of the support. Put the wood support inside the metal support bracket to obtain the exact location of this hole. Place this wood support vertically on the middle of the base and scribe around it. Chisel out a hole in the base to accommodate the support.

Before fixing the wood support in place, cut two ¾-inch-deep grooves into the bottom of the support. Glue the support in place and force glued wood wedges into the grooves to produce a tight permanent fix. Trim the ends of the wedges.

Finishing the Unit Sand and paint the whole structure according to your own taste. Push the 2¼-by-⅜-inch bolt into place through the holes in the wood and metal supports; tighten the butterfly nut.

Wood Holder A wood holder may be made to support the material to be ignited. It will have to be custom-made from copper tubing and steel rods, according to the actual size of your sealed-beam reflector. The illustration shows a wood holder fabricated for a reflector having an external diameter of 7 inches.

When your completed lighter is placed on the picnic table at your next barbecue it is guaranteed to be an attention-getter.

Bibliography

Alves, Ronald, and Charles Milligan. *Living with Energy.* New York, N.Y., Penguin Books Ltd., 1978.

Anderson, Bruce, ed. *Solar Age Catalog.* Harrisville, N.H., *Solar Age* Magazine, 1977.

Anderson, Bruce, and Michael Riordan. *The Solar Home Book.* Andover, Mass., Brick House Publishing Co., Inc., 1976.

Argus, Robert, Barbara Emanuel and Stephen Graham. *The Sun Builders.* Toronto, Renewable Energy in Canada, 1978.

Beggs, Sandra, ed. *Practical Guide to Solar Homes.* Los Altos, Ca., Bantam/Hudson, 1978.

Clark, Wilson. *Energy for Survival.* Garden City, N.Y., Anchor Press/Doubleday, 1975.

Clegg, Peter. *New Low-Cost Sources of Energy for the Home.* Charlotte, Vt., Garden Way Publishing, 1975.

Daniels, Farrington. *Direct Use of the Sun's Energy.* Westminster, Md., Ballantine Books, 1964.

de Winter, Francis. *How to Design and Build a Solar Swimming Pool Heater.* New York, N.Y., Copper Development Association, 1975.

Duffie, John, and William Beckman. *Solar Energy Thermal Processes.* New York, N.Y., John Wiley & Sons, 1974.

Eccli, Eugene. *Low-Cost Energy-Efficient Shelter for the Owner and Building.* Emmaus, Pa., Rodale Press, 1976.

Eccli, Sandy, ed. *Alternative Sources of Energy.* New York, N.Y., The Seabury Press, 1975.

Foley, Gerald. *The Energy Question.* Harmondsworth, England, Penguin Books, 1976.

Foster, William. *Build-It Book of Solar Heating Projects.* Blue Ridge Summit, Pa., Tab Books, 1977.

Foster, William. *Homeowner's Guide to Solar Heating and Cooling.* Blue Ridge Summit, Pa., Tab Books, 1976.

Halacy, Daniel S. *The Coming Age of Solar Energy.* New York, N.Y., Harper & Row, 1973.

Hand, Jackson. *Home Energy How-To.* New York, N.Y., Times Mirror Magazines, 1977.

Hand, Jackson. *Solar Heating and Cooling.* New York, N.Y., Times Mirror Magazines, 1978.

Kreider, Jan, and Frank Kreith. *Solar Heating and Cooling.* Washington, D.C., Hemisphere Publishing Corp., 1975.

Lucas, Ted. *How to Use Solar Energy in Your Home and Business.* Pasadena, Ca., Ward Richie Press, 1977.

McCullagh, James, ed. *The Solar Greenhouse Book.* Emmaus, Pa., Rodale Press, 1978.

Merrill, Richard, ed. *Energy Primer.* New York, N.Y., Dell Publishing, 1978.

Naar, Jon, and Norma Skurka. *Design for a Limited Planet.* New York, N.Y., Ballantine Books, 1976.

Nicholson, Nick. *Harvest the Sun.* Ayer's Cliff, P.Q., Ayer's Cliff Centre for Solar Research, 1978.

Nicholson, Nick. *Solar Energy Catalogue and Building Manual.* Ayer's Cliff, P.Q., Renewable Energy Publications, 1977.

Prenis, John, ed. *Energy Book 1.* Philadelphia, Pa., Running Press, 1975.

Shuttleworth, John, ed. *Handbook of Homemade Power.* New York, N.Y., Bantam Books, 1974.

Stoner, Carol. *Producing Your Own Power.* Emmaus, Pa., Rodale Press, 1974.

Vale, Robert, and Brenda Vale. *The Autonomous House.* London, Thames and Hudson, 1975.

Wade, Alex, and Neal Ewenstein. *30 Energy-Efficient Houses.* Emmaus, Pa., Rodale Press, 1977.

Watson, Donald. *Designing and Building a Solar House.* Charlotte, Vt., Garden Way Publishing, 1977.

$7.95

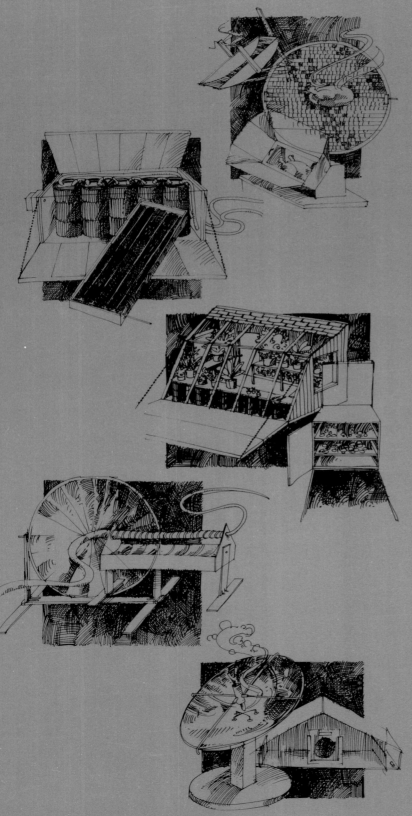

Solar hot dogs, anyone? Science teacher John Barling shows how to build a solar-powered hot dog cooker and 17 other solar "home appliances" using inexpensive recycled and new materials.

He includes both practical and demonstration projects such as:

Cooking Devices

Hot Dog Cooker
Super Hot Dog Cooker
Solar Oven
Chicken Cooker
Parabolic Concentrating Cooker

Water and Air Heaters

Water Preheating Panel
Demonstration Hot Water Heater
Barling's Barrel Bread Box Heater
Solar Swimming Pool Heater
Low-Cost Air Heater

Food-Producing Devices

Solar Cold Frame
Model Solar Greenhouse
Food Dryer

Experimental Solar Devices

Experimental Concentrating Water Heater
Fresnel Lens Concentrator
Giant Experimental Reflector

Miscellaneous Solar Devices

Glass Solar Still
Solar Wood Igniter